SAS
JUNGLE
SURVIVAL

SAS JUNGLE SURVIVAL

Barry Davies, BEM

Skyhorse Publishing

Skyhorse Publishing books may be purchased in bulk at special discounts for sales promotion, corporate gifts, fund-raising, or educational purposes. Special editions can also be created to specifications. For details, contact the Special Sales Department, Skyhorse Publishing, 307 West 36th Street, 11th Floor, New York, NY 10018 or info@skyhorsepublishing.com.

Skyhorse® and Skyhorse Publishing® are registered trademarks of Skyhorse Publishing, Inc.®, a Delaware corporation.

Visit our website at www.skyhorsepublishing.com.

10 9 8 7 6 5 4 3 2 1

Library of Congress Cataloging-in-Publication Data is available on file.

ISBN: 978-1-62087-208-6

Printed in China

Contents

When I was a young soldier, about to join the SAS, I read a book called *The Long Walk* by Slavomir Rawicz. It recounts the trials endured by six men and one woman who, after escaping from a Soviet prison camp in late 1939, walked from the Arctic Circle across the Gobi Desert and south all the way to India. After 4,000 miles and 18 months four of them, including one American, survived. Their story instilled in me a never-ending thirst for all matters relating to human powers of survival.

The jungle is, perhaps, my favourite environment in which to practise survival techniques. Jungles are divided into two types: primary and secondary. When we think of jungles we imagine a densely forested area with almost impenetrable foliage; this is what is known as a primary jungle. But jungles can also include swamps, grasslands and cultivated areas. A primary jungle can be either a tropical rain forest or a deciduous forest, depending on the types of trees and plants found growing there.

Tropical Rain Forest This typically has very tall trees whose upper branches interlock to form canopies. Underneath the top canopy there may be two or three more canopies at different levels, depending on how much light can penetrate through. The lowest canopy may be only 10m (33ft) from the ground. The effect of these canopies is to stop any sunlight from reaching the jungle floor. Undergrowth is therefore extremely limited; there are, however, extensive buttress root systems and many species of hanging vine at these levels.

As its name suggests, the rain forest has a very high rainfall, and the tropical heat produces humidity levels which at times can be seriously exhausting.

The waterlogged ground all but rules out any off-road vehicular travel, so realistically the only way to travel through this type of jungle is on foot. Due to the lack of undergrowth this is fairly easy in a tropical rain forest, especially compared to other types of forest. However, it does present its own problems: due to the dense canopy, no search-and-rescue crew will be able to see you from the air. Ground visibility is also limited to about 50m (165ft), and it is extremely easy to get disorientated and lost.

Deciduous Forest This is found in semi-tropical regions. Here the weather tends to have a marked annual season of rainfall (usually called a monsoon) and a dry season, even a drought. During the heavy rains the trees produce leaves, and when it is drier the foliage tends to die back. Unlike the tropical rain forest the trees are not so densely packed together and sunlight is able to penetrate the canopy and reach the forest floor. This encourages the development of a thick layer of undergrowth.

Travelling through a deciduous forest in the dry season is reasonably easy, and visibility is relatively unhampered. However, during the wet season, when the trees are in full leaf and the undergrowth is at its thickest, movement is extremely slow and difficult, and visibility is hampered both from above and on the ground.

Secondary Jungle This occurs on the edge of primary jungles, and appears where forest has been cleared (frequently by man, using slash-and-burn methods) and then abandoned back to nature. Where the ground has had much exposure to sunlight certain types of plants take hold and grow vigourously. These are usually weeds, grasses, thorny shrubs, ferns and bamboo.

This sort of thick, difficult vegetation, often growing to a height of 2m (6.5ft), makes any movement on foot slow and difficult. Visibility is often no more than 2-3m (6-10ft), giving a feeling of claustrophobia. Often the only way to move through the impenetrable foliage is to slash your way through with a machete.

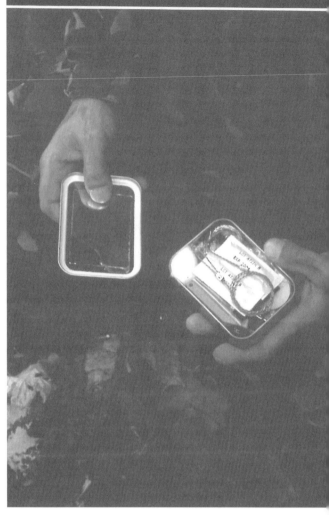

Anyone venturing into an uninhabited and potentially hostile area should carry a survival pack. The contents of the pack should be dictated by the type of terrain you are entering, and should provide the means to protect life in the event of a survival situation occurring.

Escape and survival equipment is issued as a matter of course to military pilots and Special Forces units; but the development of specialist survival equipment for civilian use has also increased dramatically over the past decade. This equipment varies from the basic items supporting such 'global' techniques as making fire, constructing shelter, and navigation, to those varying items required for survival in specific terrain and climatic conditions.

An important factor is that every item included in a survival kit has to be of real use, and its usefulness must be judged against its size and weight. Ultimately, each item must increase your chances of survival in and rescue from situations in which you may initially have no other resources apart from the clothes you stand up in.

There are few places in the wild where you won't have to deal with bugs to some degree, but this is especially true in tropical, swampy or forested areas. Every survival kit should include extra insect repellent. Those based on a solid wax stick are the best for the jungle, and give effective, long-lasting coverage.

Also adjust your medical kit so that it caters for skin rashes, snakebite, etc. In a tropical environment, or anywhere else where biting insects present a serious problem, taking plenty of mosquito netting will greatly reduce the number of bites needing treatment. If the area is extremely bad you should consider using a head-net for protection; use the type that have two hoops, top and bottom, to keep the netting away from your face.

Fire

Candle A candle will prolong the life of your matches by providing a constant flame (as long as you can protect it from wind and rain); it will help start a fire even when

Survival Kit

A survival kit could save your life. Whenever you embark upon any journey or activity where a survival situation might occur, make sure that you have a survival kit with you – and, most importantly, that it is on your person.

Each item must be evaluated for its usefulness and, ideally, its adaptability to different uses; make sure that the sole purpose of one item is not duplicated by another. Once you have decided upon your final selection, pack the items neatly in an airtight and waterproof container such as a metal tobacco tin, a screw-top cylindrical metal container, a waterproof plastic box, or a resealable polythene bag inside a sturdy canvas pouch. Whichever container you choose, once it has been packed with the relevant selection of kit for the particular conditions you face it should not be opened until needed.

The components of your survival kit should not be regarded complacently, as guaranteeing your survival without further initiative; the kit should be seen rather as a catalyst which kicks your personal survival skills into action.

the tinder is damp. Additionally, a simple candle provides light and comfort to your surroundings. Choose a candle made from 100% stearine, or tallow (solidified animal fats) – this is edible and may therefore serve as an emergency food (do not try to eat candles made of paraffin wax). The candle wax can also be used as a multi-purpose lubricant.

A flint and steel.

Flint & Steel Matches, if not protected, are easily rendered useless by wet weather, while a flint and steel will enable you to light countless fires irrespective of the conditions. The flint and steel is a robust and reliable piece of apparatus, but its usefulness is vastly improved when combined with a block of magnesium. Sparks generated by the flint will readily ignite shavings scraped from the magnesium block onto kindling materials.

Matches Ordinary kitchen matches will not be of much use unless they are made waterproof. This can easily be done by covering them completely with melted wax, or coating them with hairspray. Special windproof and waterproof matches can also be purchased; each match is sealed with a protective varnish coating, and manufactured using chemicals which will burn for around 12 seconds in the foulest of weather.

Wind and waterproof matches.

Tampon Due to the fine cotton wool used in its manufacture, the tampon has proven to be the most efficient tinder and fire-lighting aid. It works best if the white surface is blackened with charcoal or dry dirt first, as it accepts the sparks and ignites more readily. British RAF and Special Forces packs contain two tampons as standard issue. The cotton wool can also be used in medical emergencies to clean wounds.

Water

Condoms A non-lubricated, heavy duty condom makes an excellent water carrier when supported in a sock or shirtsleeve. The water must be poured in, rather than the condom being dipped into the water supply; shake the condom to stretch it as it fills up. Used in this way a

condom can hold about 1.5 litres (2.6 pints). Condoms will also protect dry tinder in wet weather; and are strong enough to make a small catapult.

Water Purification The means of water purification come in a number of different forms, from tablets to pumping devices. For inclusion in a survival kit you are best advised to choose tablets (about 50), as they are light to carry and quick and convenient to use. One small tablet will purify about one litre (1.75 pints) of water, although it will leave a strong chlorine taste. Tablets cannot clean the water or remove dirt particles, but they do make it safe to drink.

Water bottle; the newer types on the market come fitted with a built-in filtration and purification system – you simply fill them, and drink from them.

Clothing & Shelter

Needles & Pins Several different sized safety pins should be included in any survival kit. They make good closures for makeshift clothing, or can be baited as large hooks to catch fish or birds. Large sailmaker's needles, such as a Chenille No 6, have a large eye which makes threading easier, especially if the hands are cold or if you are using thread improvised from sinew. They will also be able to cope with heavier materials such as canvas, shoe leather or rawhide. Another good use for a needle is as a pointer in a makeshift compass, although it will have to be magnetized first.

Jungle Clothing

While heat and humidity are undisputed facts in the tropics, the reports of discomfort being 'unbearable' are often exaggerated. The heat would be a lot easier to bear if the moisture level was not so high; this causes constant sweating and damp clothing. This is an inconvenience, but one can learn to adapt; the jungle survivor must accept that his clothing will always be wet, either from sweat or from rain. However, problems can arise during late evening when jungle temperatures drop and damp clothing becomes chilly. It will be cold enough to warrant a fire or wrapping a blanket or some other form of covering around the body for protection.

Lightweight, loose-fitting clothes that completely cover the body are best for the jungle environment. If you arrived in your situation unexpectedly, scavenge for suitable clothing, search personal effects or improvise with any available material.

- Shirts should have long sleeves, and should button at the wrist and neck.

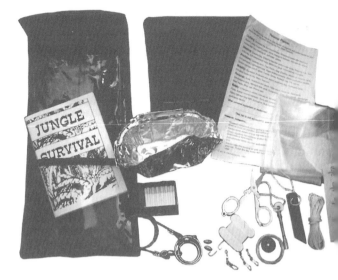

- Trouser bottoms should be gathered and tucked into the socks or boots.

- Secure all valuable survival items in pockets or around your neck on loops of string.

- Find or make a spare set of clothing for sleeping in.

- Wear a hat with a wide brim – it will help stop bits of the forest and its smaller inhabitants from dropping down your neck.

Clean Clothing Given that you will sweat constantly, your clothes will smell. Bits of jungle also get everywhere, and snagging vines are constantly ripping clothing. If you are forced to remain in the jungle for any length of time without a change of clothes then they will start to rot. The speed at which they do so will depend on how clean you keep them. If water is available then clothes should be washed every couple of days. Use any soap sparingly; a simple rinse in clean water is often enough to remove sweat and debris from the fabric. Do not beat your clothes against rocks, as this will damage buttons and zips, etc. Dry them by laying in direct sunlight; if none is available, then wring them out and wear them damp.

Footwear The best footwear is a pair of high, lace-up boots with water drainage, but these may not always be available. In this event the important factors are firmly fitting, comfortable footwear with a good sole, and a long pair of socks – with your trouser bottoms tucked inside, the latter are to protect your ankles and legs from being bitten. If you are wearing anything other than proper jungle boots you might consider making several small holes in your footwear above the sole about midway along the foot, to release any water trapped inside.

Selecting Equipment You would be astounded at the amount of rubbish untrained soldiers carry into the jungle. Prior to any operation every SAS soldier will strip his equipment down to the bare bones and carry only what is absolutely necessary. There is little point in inviting exhaustion by carrying large loads in jungle terrain, and any equipment you need to carry should be carefully selected. Remember that the jungle will provide you with most of your basic needs such as food and shelter. What you really need are your machete, compass, survival and medical kit, plus a supply of drinking water. Apart from these you should deliberate on items such as a mosquito net, spare clothing, and signalling apparatus, especially radar-reflecting balloons.

Getting Down from the Jungle Canopy If your survival situation arises from a crash-landing or a parachute exit over the jungle, there is a good chance that you will become trapped high up in the forest canopy. Parachutes will obviously get entangled in the branches; but it is also more common than most people realize for aircraft to come to rest in the canopy, given that it will simply cushion most light fixed-wing types and helicopters. If this happens, and you are still alive, your problems have only just begun. The distance from the canopy to the jungle floor can be anything up to 30m (100ft) – a fall means certain severe injury or death. While you could try climbing down, you will undoubtedly come to a point where there are no more branches yet you are still dangerously high above the ground. The only safe way is to lower yourself on a rope.

Immediate Action The ways of entering the jungle are fairly limited: either you walked or used a boat to get in, in which case you

Sleep in dry clothes

SAS soldiers have regularly operated in the world's jungles over nearly 50 years, since the Malayan Emergency campaign of the early 1950s. They have pioneered many techniques of jungle survival and warfare – methods of insertion and extraction, and long operational patrols – sometimes staying in the forest for months at a time. They have developed one completely undramatic practice which makes a big difference to comfort, and thus to efficiency.

Although dedicated to travelling as light as possible, they carry two sets of clothing with them, one for daytime use and one for night. Just before they go to sleep they change out of their wet clothing, which is normally hung under the shelter or hammock to dry out. In the morning they change from their dry clothing back into the damp set. It is an uncomfortable change, but one that guarantees a good night's sleep in dry clothes.

should be prepared – or your aircraft has made a forced landing. Military personnel who have been deliberately inserted into the jungle should have the means by which they can be extracted in an emergency. Whatever has caused your survival predicament, the same guidelines should be observed.

The immediate actions to be taken in the jungle are much like the imperatives in any other terrain: first aid, position, rendezvous and

Lowering Methods

If escaping from an aircraft trapped in the canopy, remember to remove all serviceable items, survival packs, rations, tools etc, and drop them to the ground before you lower yourself.

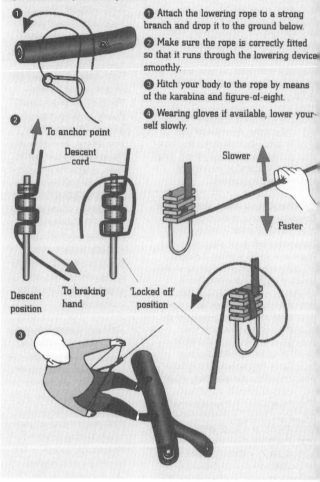

❶ Attach the lowering rope to a strong branch and drop it to the ground below.

❷ Make sure the rope is correctly fitted so that it runs through the lowering device smoothly.

❸ Hitch your body to the rope by means of the karabina and figure-of-eight.

❹ Wearing gloves if available, lower yourself slowly.

To anchor point

Descent cord

Slower

Faster

Descent position

To braking hand

'Locked off' position

contact. In the jungle first aid is particularly important, as the smallest cut or abrasion can quickly become infected. Fixing your position and rendezvousing with other survivors is also difficult due to the dense vegetation. Shouting and whistles will help at this stage. Only move when you are certain, and then move in the direction of the aircraft crash site or to a prominent feature specified by the pilot prior to jumping. Establishing communications before any emergency, or activating a personal locator beacon after the event, will greatly improve your chances of rescue. Finally, make sure that the rescue party, which will normally arrive in a helicopter, will have either a cleared landing zone or a suitable winching area.

Carry Equipment Before leaving to walk out, decide what equipment is vital to your journey and what you can easily carry.

'Para cord', a material with a score of uses in a survival situation; and hooked bungees for shelter construction.

Do take spare parachutes, ropes, fire axes, medical and survival kits, and some container which will hold fresh water. Cut up seats and cargo webbing to make improvised packs, or use a pole to carry loads between two men.

Parachute Cord Parachute cord is an extremely strong alternative to plain string, having a breaking strain of about 250 kilogrammes (550 pounds). It can be used for lashing shelter frameworks and many other necessities; and the inner strands of thinner cord also make good thread for sewing or fishing lines. A survival kit should contain a minimum of 15 metres (50 feet).

A stainless steel wire saw will cut through most materials, including steel.

Razor Blades Hard-backed razor blades make useful cutting tools, for gutting fish, cutting sinew, or when making a weapon. Despite its small size, if used with care the blade will continue to cut for up to a month. Its life can be prolonged by not trying to cut materials which are obviously beyond its capabilities.

Wire Saw A good saw, similar to those issued to the military, is made of eight strands of stainless steel wire; it is capable of cutting through wood, bone, plastics, even metal. A wire saw can be used when cutting timber to make a shelter, and to saw precise notches when constructing traps and snares. (Due to the friction a wire saw may overheat – saw slowly so as to avoid this.) The saw can even be used as a snare itself, by passing the smaller ring at one end through the larger ring at the other to form a running noose.

Navigation & Signalling

Air Marker Panels Air marker panels can be made from any light-weight fluorescent material, although orange is the standard recognizable colour. It is advisable to carry a sheet two metres (at least six feet) square, which can either be folded into different shapes (see under Signalling, page 149), or split into three 30cm (12in) wide strips. Do not split the sheet until there is an absolute need to do so – e.g. you have spotted a search aircraft – as the whole sheet is useful for other functions such as a makeshift shelter.

Compass A compass provides the means to establish direction and position, the two vital elements if you are forced to travel or need to give your co-ordinates during rescue. A small button compass is designed primarily as an emergency escape and evasion tool for the military, its use being restricted to direction-finding only.

The liquid-filled 'Silva'-type compass is more commonly associated with navigation by map, and together these offer the means of precise position-finding.

Flares & Smoke A wide variety of signal flares and smoke canisters are available on the market. If you decide to add them to your survival kit you would do best to choose a standard flare pack containing a

'Silva'- type compass.

launch pistol and nine different coloured flares. Although it is a good idea to know which colour is traditionally associated with which intended signal, firing any colour will attract attention. When firing the flares take great care that the launch pistol is aimed sky-wards. In a life-threatening emergency flares can be used to start a fire.

Global Positioning System (GPS) GPS is relatively new to the survival market, but its popularity is growing. This state-of-the-art instrument is a navigational aid capable of plotting your precise position on the surface of the Earth. This is obviously of particular value in the Arctic, where a normal compass can become erratic. However, GPS has certain drawbacks in most pure survival situations; away from any other power source it relies on batteries, making its usefulness short-lived.

The Global Positioning System (GPS) utilizing the network of navigation satellites in Earth orbit is of limited value in most survival situations; although it may allow initial positioning, in the absence of a power source its batteries do not last long.

Heliograph Modern heliographs are small, light, and easy to use. They operate by reflecting the sun's rays precisely towards aircraft or other rescuers. On a clear, sunny day their reflection can be spotted up to 30 kilometres (18.5 miles) away. It is a good idea to familiarize yourself with the operating instructions prior to any rescue attempt.

Radar-Reflective Balloon Radar-reflective balloons are not new to survival, but in recent years they have improved dramatically. The principle is to inflate a balloon made from a special foil which can be detected by search-and-rescue radar from ranges of up to 38 kilometres (24 miles). Some are tethered to a length of line and flown like a kite, while others are inflated by gas; the latter will stay aloft for up to five days even in strong winds.

Strobe Designed for military rescue situations, the strobe is a bright blue light which flashes with great intensity and can be seen many miles away. These are perfect for location at night or in the darkness of the Arctic winter. Although the strobe is waterproof, in cold conditions it should be kept close to the body to preserve the battery strength. The strobe should only be operated when the sound of a rescue aircraft is positively identified.

Survival Radio Although there are a vast number of survival radios on the market, some are limited in range and capabilities. If your work or pastimes often take you into isolated areas then you are well advised to carry a radio telephone which is capable of world-wide communications. Many surviving parties have been successfully rescued by telephoning the emergency services directly. Mobile phones will only work where there is a network, but the introduction of global satellite phones will improve things.

Watch Although not a direct part of your survival kit, a watch can be an excellent navigational aid – providing that it is of the analogue type, i.e. not digital.

Whistle Modern survival whistles are compact and can have a range of up to 1000 metres (5/8 mile) on a clear day. although this is of lesser benefit in the jungle where sound does not carry well.

Food

Emergency Food Food is plentiful in the jungle and is not an immediate requirement in a survival situation, as the body can do without solids for several weeks before it starts to deteriorate. Any food pack should be kept to the minimum: two fuel tablets, two tea or coffee sachets, sugar, etc.

All survival packs should contain basic fishing kit, which has a high usefulness-for-weight value.

Meat stock cubes contain salt and flavouring, making excellent hot drinks. They can add flavour to plants and food from the wild.

Fishing Equipment A survival fishing kit should consist of the following basic components: five hooks (size 14 or 16), approximately 30m (100ft) of line, 10 iron or brass weights, and swivels. A float can be made from a cork. If there is room, include a luminous lure, and small fishing net.

Snares Purpose-manufactured snares work best, but if you cannot get these then carry at least 5 metres (16.5 feet) of brass wire to make snares. Next to a rifle, snares are perhaps the most effective way of catching game. Brass wire can also be used for fishing traces, and when building shelters or improvised packs.

Knife

In the jungle a large knife or machete is essential, but if you have crash landed unprepared in a jungle area you must make do with whatever is to hand.

Along with your survival kit you should select a good knife. For many reasons this may be the most important item you carry.

Pocket knives range from the simplest single-bladed type to multi-bladed, multi-function knives. Whichever knife you choose, always carry it on your person. Single-bladed knives offer little more than a simple cutting tool. You will be better off with a multi-function, 'Swiss Army' type. This provides the survivor with a versatile range of tools, including scissors, saw and screwdriver.

Survival knives tend to be much larger than pocket knives and are usually carried in their own sheath. Most of the better knives have a sharpening stone in a pocket on the sheath, and many have a hollow handle or a pouch on the sheath in which a basic survival kit can be carried, although the nature and number of items in the kit will depend on the price.

A multi-purpose 'Swiss Army' type knife.

The next step from the 'Swiss Army' knife is the multi-function tool such as the Leatherman or Gerber, whose increased size and length gives extra leverage and weight to its applications.

Multi-functional tools serve endless applications, from shelter building to making improvised clothing and travel gear. In a survival situation the multi-function tool is likely to prove more productive than an ordinary knife. The better-known names such as Leatherman and Gerber are well made and should last a lifetime. Most types include pliers, wire cutters, cutting blade, saw, screwdrivers and files. Because of its importance the tool should be attached to the body by a length of cord to prevent loss.

Jungle Survival Tool Any planned trip into the jungle will warrant taking a large knife or machete. This will be required to clear vegetation from your path and to construct a camp site. The Rockwell 55/57 has been developed after years of research in

Keeping a Sharp Edge

The knife is a vital survival aid; do not misuse it by throwing it into the ground or at trees. Keep it clean, and know where it is at all times.

A knife with a blunt edge is nothing more than a useless piece of steel. Granite or dark, hard sandstone are best for sharpening a blade. Find a flat piece the size of an open palm; rubbing two rocks together will produce an even surface. Wet the stone surface and work the blade edge over it with a smooth action, always working the blade away from you across the stone. At first use a clockwise circular motion over the surface, then an anticlockwise motion. Learning to sharpen a blade is a skill that can only be achieved through practice.

Grinding the blade at the correct angle will produce a long-lasting cutting edge. If your intended travel will involve a lot of cutting then you will be better off carrying a sharpening stone or steel with you.

The new jungle machete, developed by former SAS soldiers, is a major improvement on the former issue.

conjunction with former SAS soldiers, and is undoubtedly the most suitable tool for the job. It has a heavy chopping blade with which it is possible to fell large trees, but with a finer edge for skinning and food preparation. The broad, flat point makes an excellent digging tool, or can be used for boring holes. Engineered from 440a stainless steel, with a tough plastic handle, this tool will perform a multitude of tasks and withstand the most extreme treatment.

Survival Medical Pack

Knowledge of even basic first aid skills is a useful and valuable accomplishment in everyday life, but in a survival situation these skills take on immeasurable importance. Even when medical training and equipment are non-existent, it is always possible to save life if the priorities of first aid are administered.

It is important to put together a small emergency medical kit, based on your own personal medical skills. Obviously, if you are not trained as a medic, you should only include basic items, as detailed below.

Antihistamine Cream Antihistamine cream will soothe the severe irritation that insect bites or allergies can cause. Antihistamine tablets can be carried as an alternative, but beware – some cause drowsiness.

Antiseptic Potassium Permanganate crystals (see below) are easy to carry and provide an all round sterilizing agent, antiseptic and anti-fungal agent. A tube of general purpose antiseptic cream is also very handy.

Aspirin Aspirin will relieve mild pain and headaches and reduce a fever. Carry a strip of about a dozen soluble aspirin tablets.

Dressings Include at least one large wound dressing in your medical kit. Always have it ready for immediate use. Note – the inside of a wound dressing contains cotton wool which makes excellent tinder, so make sure to retain all used dressings.

Electrolyte Drinks Most survivors will inevitably suffer from dehydration; this can occur in both hot and cold climates, and is mostly attributed to diarrhoea. While replacing fluid loss is the priority, body salts and minerals can also be replaced by adding an electrolyte drink.

Magnifying Glass In survival situations a magnifying glass is traditionally associated with fire-lighting by focusing the sun's rays on dry tinder; however, it is also useful for finding hard-to-see objects such as splinters and thorns. Short, sharp burns are also effective for removing leeches and ticks from the body.

Mosquito Repellent The chances of contracting malaria and other mosquito-borne diseases can be reduced if the correct precautions are taken. Anti-malarial tablets, as prescribed by a doctor, need to be taken; but it is just as important to deter the insects from biting you in the first place, so it is recommended that you include a mosquito/insect repellent in your kit.

Plasters Carry various sizes and shapes of waterproof plasters. Larger plasters are best, as they can always be cut down if necessary. Keep your plasters together in a waterproof sachet.

Potassium Permanganate This crystalline chemical has many uses, and is carried in military survival kits. If mixed with a glycol-based substance such as antifreeze it can be used to light a fire. A small amount added to water will make a sterilizing mouthwash, and a concentrated mixture can be used to treat fungal diseases.

Surgical Blades Two surgical blades take up little space and are best left in their protective sterile wrapping. In use they can either be held between the fingers, or a handle can be fashioned from a small stick. Do not discard used surgical blades; sterilize by boiling, and re-wrap. A blade no longer viable for surgery makes an excellent arrowhead.

Salt Salt is essential when travelling in tropical climates. Carry a small amount to make sure that the salt balance in the body is maintained. Try to reserve this resource for medical uses and refrain from using it for cooking. Salt water is helpful in treating fungal infections.

Suture Plasters If you are unable to administer stitches, butterfly sutures will prove successful in closing small wounds.

The first task for any survivors is to establish the priorities for treatment of the injured. Casualties are usually sorted into categories. Those who require urgent assistance to prevent immediate death must obviously be given priority.

Anyone suffering from an asphyxia disorder will need immediate attention. Shock caused by major injuries and severe haemorrhaging must be assessed; after a major disaster many may be hopelessly injured and thus cannot qualify for immediate assistance. The task is to identify the injury and establish how long the casualty will live without assistance; and to decide if any assistance that can be given will prove beneficial.

Breathing

Check a casualty's breathing by placing your ear close to the nose and mouth and looking down over the chest and abdomen.

If they are breathing you should be able to both feel and hear the flow of air, and to see chest and abdominal movement. If these signs of breathing are absent, immediate action must be taken.

First make sure that the airway is clear:

Tilt the casualty's head back gently while lifting the chin with the other hand. Doing this will automatically open the airway, and will also lift the tongue from the back of the throat so that it will not cause an obstruction. Supporting the head in the tilted position with a hand on the forehead, check inside the mouth for any blockage, e.g. dentures, vomit, etc. Gently remove blockages, without touching the back of the throat, as this may cause throat tissue swelling.

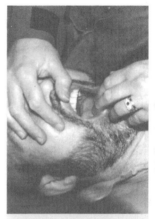

Check that the airway is clear.

In many cases these actions alone may be enough to enable the casualty to breathe again. If this is the case, and they also have a pulse, then place them into the recovery position and maintain a periodic check on their condition.

Any visible injury to the front or back of the head may also indicate that the casualty has damaged his neck or spine. In such cases maintaining an open airway will still be a priority over their other injuries. However, it is recommended that a collar or head support be improvised in order to keep the head properly positioned.

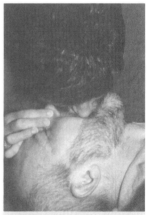

Mouth-to-mouth resuscitation.

Artificial respiration If the casualty is still not breathing, you must get oxygen into their body. This can be achieved through mouth-to-mouth resuscitation, as the air we exhale still contains 75% oxygen.

LIFESAVER

The human body needs a constant supply of oxygen to function. If we stop breathing even for a few minutes the brain will start to sustain damage; the longer we go without air, the greater the damage. If casualties are unconscious, choking, or having trouble breathing, then they must be treated urgently.

In the case of an unconscious casualty, check for breathing and also for a pulse. If one or both are undetectable then emergency treatment must be given immediately. Urgent assistance must also be given to anyone who is choking or showing other obvious signs of breathing difficulties.

With the casualty's head still tilted back so that the airway is clear, pinch his nose to prevent air loss.

Breathe in deeply, and then seal your lips over the casualty's mouth. Gently blow into their mouth and watch for the chest to expand. It will take about two seconds for the chest to expand to its maximum capacity. Move your mouth away and wait for the chest to fall fully.

This should be repeated nine times before checking that the casualty's heart is still beating by feeling the carotid pulse point in the neck. It is no use providing the patient with oxygen if their heart is unable to pump blood to the necessary organs. If the heart has stopped, chest compressions (see below) must be administered.

In cases where mouth-to-mouth resuscitation is impossible or undesirable, e.g. when there is a serious lower jaw injury, mouth-to-nose ventilation may be carried out instead, but making sure that the mouth is firmly sealed first.

Artificial respiration should be carried out until the casualty is once more able to breathe unaided. Once the breathing rate is steady, place the casualty in the recovery position and monitor their condition every three minutes.

Chest Compression If the heart has stopped, it must be artificially pumped so that the oxygen carried by the blood can reach the vital organs. To do this a technique called chest compression is used.

Artificial respiration when there are two people available.

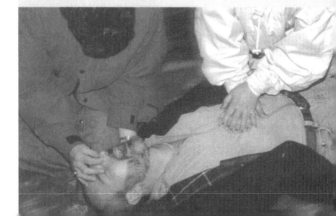

Before beginning this procedure it is vital to make sure that a pulse is entirely absent. If the heart is still beating, however weakly, then chest compression will cause damage.

Place the casualty flat on his back on a firm surface. Kneel beside him and locate the bottom of his breastbone – this is found where the bottom two ribs meet. Place the heel of one of your hands about three fingers' width up from this point; place your other hand on top of this, and interlock the fingers.

Lean forward over the casualty, making sure that your elbows are rigid and that your weight is pressing vertically on the casualty's chest. The breastbone should be depressed by about 4-5cm (2 inches). Release the pressure by leaning back, but without removing your hands. Chest compressions should be repeated at a rate of about 80 per minute, pausing for a pulse check every 15 compressions.

Generally, if the heart has stopped then breathing will also have stopped. In order for the casualty to have a chance of survival both artificial ventilation and chest compressions will have to be performed at the same time. If you are on your own, the correct procedure is to first give the casualty two assisted breaths, followed by 15 chest compressions. Continue with this cycle for one minute before checking on heartbeat and breathing.

If neither is present, continue with the alternated breaths/chest compressions until either the casualty's heartbeat is restored, help arrives, or you become too exhausted to continue.

Resuscitation by two people
If a second person is present and able to help, one should assist the casualty's breathing while the other manipulates the chest compressions. To begin with, four

While assessing your priorities, keep these rules in mind

❶ Do not panic, no matter how serious the situation. Panic means that you will think less clearly. Take several deep breaths to calm yourself.

❷ Each casualty's injuries will need to be assessed. You will need to use all your senses: ask (if the casualty is conscious); look (and if possible feel over the body for broken bones, blood etc); listen; smell; think – and act.

❸ Conscious casualties are an important source of information. Ask them to describe their symptoms and what they feel may be wrong with them.

❹ Avoid taking any action that will put you in danger. If you become injured, you will not be able to help anyone else.

❺ Boost the morale of your casualties. Offer bcomfort, reassurance and encouragement, thus building their mental strength and will to live.

❻ Get uninjured survivors to help you. Ask (out of earshot of any injured person) if anyone else has any medical experience.

❼ Separate those who are saveable from those who are not.

Attempt to remove any obstruction by slapping between the shoulder blades.

assisted breaths should be given followed by five chest compressions. Subsequently, the correct procedure is to give one assisted breath for every five chest compressions. There should be no pause between the end of the chest compressions and the beginning of the assisted breath.

After one minute check for pulse and breathing; if neither is present continue the alternating breath/compression cycle and check every three minutes. Continue until either heartbeat and breathing are re-established, help arrives, or both helpers become too exhausted to continue. If heartbeat and breathing do return, check for any other injuries and place the casualty in the recovery position.

Choking Choking requires immediate assistance, as the airway is blocked and therefore no air is getting through to the lungs. Choking can be recognized by the casualty suddenly being unable to breathe or speak, grabbing at their throat or their skin turning pale blue.

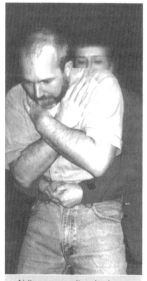

Aiding a casualty who is choking.

The first priority is to remove whatever is causing the blockage in the windpipe. If the casualty is conscious, get them to cough it up. If this does not work, check the mouth to see if the object can be cleared with a finger. If not, bend the casualty as far forward as possible, preferably so that the head is below the level of the lungs. Give five sharp slaps between the shoulder blades with the heel of the hand, and check to see if the obstruction has been dislodged. This is usually enough to remove the object, but if it does not work you will have to try to clear it by using abdominal thrusts.

To do this, stand behind the casualty and put your arms around him. Ball one of your fists and lock

The Recovery Position

An unconscious casualty with a regular heartbeat and who is breathing normally while showing no sign of serious injury should be placed in the recovery position. In this position the head is slightly lower than the body, thus preventing the tongue from blocking the airway and allowing any liquids such as blood or vomit to drain freely from the mouth.

Kneel to one side of the casualty and turn his head towards you. Straighten the nearest arm alongside the body, with the other folded across the chest. Cross the ankles and roll the casualty towards you. Gently bend the upper arm and leg so that they safely maintain the body in the position. Keep the head, neck and back in a straight line.

If the casualty has spinal injuries or wounds the position may have to be slightly modified. In such cases use improvised padding, such as towels or rolled clothing, for extra support.

it in place with the palm of your other hand, making sure that one thumb is pressing into the abdomen. Pull your hands sharply inwards under the casualty's ribs. Repeat up to four more times before checking whether the object has been expelled. If this does not succeed at first, give five more back slaps and then five more abdominal thrusts. Keep trying until the object becomes dislodged.

Choking when unconscious If the choking casualty becomes unconscious, first lie him on his side with his abdomen supported against your knee, and give four to five back slaps.

If this does not dislodge the object, turn the casualty onto his back, kneel astride him, and perform abdominal thrusts. To do this, locate the heel of one hand just below the ribcage and cover it with the other hand. Press sharply inwards and upwards with the heel of the hand, up to five times. Check in the mouth to see if the object has been expelled. Continue alternating back slaps with abdominal thrusts until the obstruction is removed. If the casualty begins to breathe normally, place him in the recovery position and check breathing and pulse rates every three minutes.

If breathing does not recommence and/or there is no pulse, start assisted breathing and, if necessary, chest compressions.

Self-help when choking If you find that you are alone and choking, find something like the back of a chair or a tree trunk, and push it inwards and upwards into your abdomen to expel the air and, hopefully, the blockage. You could also attempt to use your own hands made into a fist to achieve the same effect.

Pressure Points

Indirect pressure utilizes pressure points. These are found where arteries cross bones near the skin's surface. For survival purposes, concentrate on the four points which flow to each limb.

❶ The pressure points in the arm are found down the centre of the inner side of the upper arm, on the brachial arteries.

❷ The main pressure point in the leg is on the femoral artery. The pressure point for this artery is in the middle of the groin. It is often easier to locate if the knee is bent so as to create the groin crease. Press firmly at this point against the bones of the pelvis.

❸ Locate the pressure point and, placing the thumb or fingers on it, apply enough pressure to flatten the artery against the bone. This should stop the blood flow.

❹ Pressure must not be kept on for any longer than ten minutes, or else other healthy tissue will be damaged through lack of blood. While using indirect pressure the wound may be dressed more effectively; however, do not use a tourniquet, as this may cause tissue damage.

Bleeding

Once breathing and circulation are restored the next priority is bleeding. Bleeding may be external or internal. Internal bleeding is almost impossible to treat with first aid, but external bleeding can be controlled.

Wounds present two main problems. Firstly, extensive bleeding can cause shock to develop, and will, if not controlled, lead to death. Secondly, any break in the skin will let infection in, so it is imperative that the wound site be kept clean. There are three procedures to help stop the bleeding:

Direct Pressure Use a sterile dressing if you have one; if not, find a clean piece of cloth. Place the dressing on the wound and press on it gently but firmly. If you have no dressing available then you may have to use your hand, but bear in mind the dangers of infecting the wound. Use only dressings that are large enough to

cover both the wound and part of the surrounding area. It is possible that the first dressing will become soaked through with blood. If this happens, lay a second dressing over the first and, if necessary, a third over the second.

Tying a bandage around the wound and dressings will keep the dressings in place with a continued firm pressure. It is important that the bandage is not tied too tight, as this will restrict the flow of blood to the whole area.

Some large wounds will tend to gape. If you have suitable dressings you may use these to bring the edges of the wound together; otherwise you may have to use your hand. Blood flow from a large wound may be stopped by applying firm pressure with a pad of dressings, to the site of the greatest bleeding.

Using pressure on the wound helps the body's own mechanisms to slow and finally to stop the bleeding. The damaged ends of blood vessels will shrink and start to retract in order to slow down the blood loss. Clotting agents are released so that the blood begins to thicken, and eventually forms a plug over the wound.

Applying a Tourniquet

1 Use a loop above the wound secured with a square knot.

2 Insert a strong stick or similar under the loop to act as a tightening device.

3 Twist the stick, tightening sufficiently to stop bleeding.

4 Secure the stick to prevent the tourniquet becoming loose.

Sometimes these mechanisms will be enough to stop the bleeding on their own. However, the casualty may still be in danger of going into shock. It is therefore vital that they rest; and reassurance, too, is important – if the casualty is anxious it will only serve to raise his heart rate and blood pressure. An injured limb should be elevated above the level of the heart, as long as it is comfortable for the casualty and not liable to make any other injury worse. This elevation reduces the flow of blood to the damaged area and helps the veins to drain blood away, reducing blood loss. The elevated limb should be supported if possible, either by you or by padding.

Indirect Pressure If, due to the severity of the bleeding, the techniques described above do not work, then indirect pressure should be tried. However, this only works on arterial bleeding, so it is important to identify what type of bleed you are dealing with.

Arterial bleeding takes place from vessels which are carrying filtered and oxygenated blood away from the heart and lungs. It has

Fractures

1 Complicated fracture where broken bone has damaged blood vessel.

2 Open fracture where bone is exposed.

3 Closed fracture where bone is not exposed.

4 Secure fractured limb above and below the knee and at the ankle.

5 Use padding with foot injuries. Elevating the foot reduces swelling.

6 Support a broken arm with a splint improvised from rolled up newspaper. Never use metal splints in a cold climate. Immobilize the arm with a sling to speed recovery and avoid further injury.

no impurities and is therefore bright red. It will also spurt out of the wound in time with the heartbeat.

Venous bleeding takes place from vessels which are carrying blood full of impurities away from the tissues towards the heart and lungs to be filtered and re-oxygenated. As venous blood is low in oxygen it is dark red in colour. It runs steadily or gushes from a wound at a steady rate.

Tourniquets The aim of first aid is to save life. If the damage to a limb is so severe that it plainly requires amputation, or if part of the limb is missing, and direct pressure will not stop the bleeding, then you may need to employ the third procedure by applying a tourniquet.

The tourniquet can be made from whatever cloth is at hand, but avoid any thin material that will cut into the flesh. Place it around the extremity, between the wound and the heart, 5 to 10cm (2-4ins) above the wound site.

Never place it directly over the wound or a fracture. Use a stick as a handle to tighten the tourniquet, but tighten it only enough to stop blood flow. Clean and bandage the wound.

The tourniquet must be slowly released every 10-15 minutes for a period of 1-2 minutes; however, you should continue to apply direct pressure at all times. It must be stressed that applying a tourniquet to prevent blood flow is a dangerous procedure, and should only be attempted when all else has failed.

Fractures

Fractures normally occur during an accident in which a body has stumbled unrestrained or has been hit by a flying object. Not all fractures are readily apparent, but a casualty may have a bone frac-

A skull fracture or concussion must be suspected if any or all of the following symptoms are present:

1. An obvious head wound, a bruise or a soft or depressed area on the scalp.

2. Unconsciousness, even for a short period of time.

3. Clear or watery blood coming from the ears or nose.

4. Blood in the white of the eye.

5. The pupils of the eyes are unequal or unresponsive.

6. A steady deterioration in responsiveness to external stimuli.

ture if he has difficulty in moving a particular part of the body normally. The reason for a fracture is fairly evident, and it is followed by a sharp increase in pain when movement of the affected part is attempted. Pronounced swelling, bruising, distortion and tenderness at the site of the injury are also good indicators of a fracture. An injured limb may look deformed or shortened, and a distinctive grating sound may be heard while attempting to move the limb. Signs of shock may be evident, especially if the injury is to the ribcage, pelvis or thighbone. The casualty may also have felt or heard the bone break.

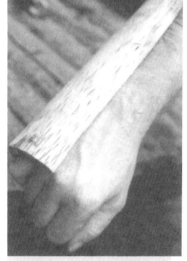

Splints can be improvised with composite materials as well as single objects.

In a survival situation the scope for treatment of a fracture is limited to immobilization of the injured part.

Splints should be applied before the casualty is moved unless there is some form of imminent danger which requires immediate evacuation. If conscious, fracture casualties will be experiencing pain, so handle them with the greatest of care so as not to cause increased distress. If the fracture has also caused a wound, this must be treated and stabilized before any splints are fitted.

Splints Suitable splints can be improvised from small branches, sticks, or suitable pieces of equipment; rolled clothing or bedding can also be used in an emergency. Make sure that the splint is padded and that it supports the joints both above and below the fracture. In the case of a leg fracture, if no suitable substitute for a splint can be found in your environment then immobilize the injured leg by tying it to the good leg instead.

Sometimes a fractured limb may become twisted, shortened or bent in such a way that immobilization proves impossible. Gentle traction to re-align the limb can be used as long as the casualty can tolerate the pain. Pull gently in a straight line with the bone until the limb has been straightened. If this is done properly the casualty may find that the pain and any bleeding at the site of the fracture are significantly reduced.

Once you have done all you can to straighten the limb, apply the splints. If possible, elevate and support the fractured limb as this will help to reduce both any swelling and the danger of the casualty going into shock.

Make sure that the casualty receives plenty of rest.

Open Wounds

Cleaning Open Wounds The purpose of washing a wound is to remove as much bacteria as possible, thus giving the body's own defensive system the best chance of finishing the job. All exposed wounds, no matter how small, need to be cleaned. This is best done with water which has been sterilized by boiling, but clean, pure drinking water will suffice if boiling is not possible.

Deeper wounds can be washed out more efficiently by making some form of irrigation device which will deliver a strong jet of water into the wound. A small plastic bottle or a polythene bag can be pierced with a pinhole so that the water jets out when it is squeezed.

Adding a very small amount of soap or potassium permanganate to the water will assist in flushing out the wound. The amount of potassium permanganate crystals added should be barely enough to tint a pint of water; similarly, only enough soap should be added to barely cloud the water. If in doubt, err on the side of weakness.

Debris and Foreign Bodies Before starting any cleaning or irrigation, open the wound to its fullest extent and examine for debris – bits of clothing, glass, dirt, or any other foreign body which may have been forced into the wound at the time of injury. If these are small and not deeply impacted, remove them; if no properly sterile instruments are available, wash your hands with soap and water

LIFESAVER
No matter what your situation, if you intend to handle open wounds or burns – whether on yourself or other casualties – you should reduce the risk of further infection by sterilizing your hands. Wash them with water, snow, alcohol, or anything that will disinfect them.

and use your fingers. (Instruments and wound dressings can be sterilized by boiling for five minutes.) Once the wound is open and foreign bodies have been removed, scrub it briskly while irrigating at the same time – this is a job best done by two people. Work quickly, as this will be very painful for the casualty. Once finished, apply a clean sterile dressing, and arrest any fresh bleeding by direct pressure. Check the wound on a daily basis.

Unless they are life-threatening, larger foreign bodies deeply impacted should be left in place, as pulling at them may cause more serious damage. Control the bleeding by direct pressure, squeezing the wound along the line of the foreign body. Next, form a padded ring which will fit neatly over the protruding object, and secure it with a dressing.

Sucking wounds If air is allowed to enter the lungs from puncture wounds to the chest or back then a sucking wound will develop. Always check for sucking wounds if missiles or debris of any form have penetrated deeply, or if a rib is protruding from the chest or back. The lung on the affected side will collapse, and as the casualty breathes in so the sucked air will also impair the efficiency of the good lung. If the condition goes untreated the result will be a lack of oxygen reaching the bloodstream, which could cause asphyxia.

If a sucking wound is suspected, immediately cover the area with your hand. Support the casualty in a lop-sided sitting position with the functioning lung uppermost. Cover the wound with a clean dressing and place a plastic sheet over the top so that the plastic overlaps the dressing and wound; tape it down so as to form an airtight seal. If a foreign body is present in the wound, do not remove it, but pack with a ring as described above and fit an airtight seal.

Dislocations

Dislocations are caused when bone joints become separated and get out of alignment. This can be extremely painful, as the nerves and blood flow are affected. The best way to relieve this pain is to re-align the joint as quickly as possible. Although this is a simple process the joint will be swollen and extremely

tender and the limb will suffer a lack of mobility.

Dislocations are treated by reduction or 'setting' the bones back into their proper position. There are two basic methods available to the survivor, depending upon whether they are alone or not. In either case the appropriate action should be taken as quickly after the dislocation as possible. Both, if successful, will bring about a lessening of pain and restoration of the circulation. Once reduction is completed the limb should be immobilized, using splints if possible, and allowed to recover.

Using a weight to assist reduction of a dislocation when there is nobody else available to help.

Use a well-padded splint above and below the injury site. Always check the circulation below the dislocation after completing the splint. Remove the splint after a week and start gentle exercises until the limb is fully functional.

Unassisted reduction The lone survivor will need to improvise some form of weight, e.g. a large rock or log, to which they can attach a cord from the limb. The idea is to stretch the limb slightly by countering against the weight, and aligning it back into place.

The procedure requires the body and/or limb to be rotated in order to set the joint while at the same time comparing it to the joint on the opposite side. All movement must be kept to a minimum, yet must be positive rather than hesitant.

The procedure should be performed lying down if possible, as it is extremely painful and the manoeuvre will require a great deal of will-power on the part of the survivor.

Assisted reduction The same basic procedure of stretching and re-aligning the limb is followed, but it has the advantage of being more often successful, since manipulation is usually more positive and precise when the casualty does not have to deliberately inflict pain on himself. Where possible one person should hold the casualty in a comfortable position while a second manipulates the limb into alignment. Again, this procedure is best done with the casualty lying down.

Concussion & Skull Fractures Skull fractures and concussion are also common after major accidents.

Concussion is a temporary disturbance of the brain, normally due to a severe blow or shaking. If conscious, the casualty should be made to lie down with their head and shoulders supported. If unconscious, make sure that they are breathing and have a pulse – if not, carry out artificial ventilation and chest compressions immediately.

If the casualty is unconscious but the breathing and pulse are normal, turn them into the recovery position and maintain a close check on their vital signs.

In either case, make sure that the casualty is kept warm and quiet and handled carefully. Apply a light padding to the injured area and hold it in place with a dressing. If blood is being discharged from an ear, lightly cover it but do not block it. Concussion is normally only a temporary disturbance from which the chances of recovery are good.

Burns

Naked flames, boiling water, electrical devices, friction, acid, liquid oxygen, freezing metal and the sun all cause skin burns. The severity of the burn and the amount of body area affected will determine the casualty's survival chances.

Cooling Burns caused by naked flame should be cooled immediately to limit the damage caused by heat to the skin tissues. Either pour cold water slowly over the affected part, or immerse it totally in clean cold water.

This should continue for at least ten minutes to stop further tissue damage and to reduce pain and swelling.

Dressing Once the burn has been cooled, a dressing should be applied immediately to limit the possibility of it becoming infected. Do not attempt to remove any charred fibres that have stuck to the burn, but remove any restrictive clothing around the site to prevent further swelling. The dressing should be sterile and made of a non-fluffy material. Avoid adhesive dressings, which will only aggravate the injury and cause more damage.

In a survival situation sterilization of cloths, bandages and dressings can be achieved by scorching the cloth with a candle, as this will kill most bacteria. Do not be tempted to burst any burn blisters which form, as these provide a protective layer. A solution containing tannic acid derived from boiling oak or beech bark can be used to clean the burn; make sure that any such concoction has cooled before using it. If polythene bags are available they can be used to cover the burnt limb and help stop further infection.

To reduce the possibility of shock setting in, lay burn casualties down and keep them warm and comforted. If the casualty is unconscious, turn him over into the recovery position and monitor his breathing and pulse closely.

Burns

1. Cool the burnt area by immersion in cold clean water, or fresh snow.

2. Protect hands and feet from further infection with a sealed polythene bag.

3. Do not use adhesive or fluffy dressings.

4. Do not break blisters or remove loose skin.

5. Do not apply ointment, oils or fats to the burn.

Survival Medicine

After the immediate priorities for first aid have been identified and acted upon, survivors in remote and uninhabited regions will face the need to monitor and safeguard their health under challenging conditions, perhaps for quite long periods.

In a survival situation it is obviously important to maintain health while awaiting rescue – and this is absolutely central to preserving the survivors' option to make an attempt at travelling across country themselves.

Fear The greatest danger for those who have never been in the jungle before is fear. Strange and startling animal noises, large biting insects, the dangers of sickness, and the sheer claustrophobia of the dim surroundings can make the jungle environment seem like a hostile place. While the jungle is noisy, smelly and teeming with animal life, it is not as dangerous as most people perceive; indeed, in many ways these forests are places of great beauty. The secret is to go with the jungle – if you try to fight it, it will fight back, making your life even more miserable and lessening your survival chances.

The forest will provide food, water, and materials to construct a shelter or make weapons with which to protect yourself. Most of the hazards are avoidable. If you are uninjured, fairly fit, and have a little common sense, you have no reason to fear the jungle.

Heat Heatstroke or heat exhaustion is common in the jungle. If the victim shows signs of becoming weak or giddy, and the skin is clammy and cold, you must suspect heatstroke. Find a shaded spot and let the victim rest, loosening clothing to circulate the air over the skin. Pour cold water over the head and neck, and give water with a little salt in to drink. Rest until recovered; take things slowly to prevent recurrence.

Personal Hygiene When ill, weakened or injured, the first priority is not medicine but a good standard of personal hygiene.

Proper hygiene, care in preparation of food and drink, waste disposal, insect control (and, of course, prior immunization) will greatly reduce the number of diseases and infestations to which the survivor may fall victim. The importance of prevention of disease during any survival situation cannot be overstated, and physical hygiene is important if you wish to protect yourself.

Wash daily with warm water and soap. Sponge the face, armpits, crotch and feet at least once a day. Regular washing, especially after defecation, is a necessity no matter how bad you are feeling.

Immunization

Vaccinations are available against many diseases which may be threats in a survival situation, including typhoid, paratyphoid, yellow fever, typhus, tetanus, cholera and hepatitis. It is essential to obtain as many vaccinations as possible, making sure immunization records are kept up to date. Prior to visiting any foreign country you are advised to seek current medical advice and take extra immunizations and precautions. Make sure that you carry a good supply of anti-malarial tablets where necessary.

Eat as often as you can to keep up strength, but stick to simple, easily digested and nourishing foods – a vegetable soup is ideal. Make sure that your water is pure and that you wash food before cooking it. Raw food should be avoided as it is not only harder to digest in most cases, it is also a possible source of contamination. Make sure that your liquid intake is sufficient – water or a herbal tea is the best drink for a body in a weakened state. Although these points are common sense, they will aid the body to heal itself or to fight off an illness or infection.

Underclothing collects dirt and sweat, so keep it as dry and clean as possible, especially if you are unable to change it regularly.

Lice & Ticks Along with ticks, lice are also carriers of typhus, which is transmitted through their faeces.

Clothing should be checked regularly, and if any of these pests are discovered they should be removed either with a delousing powder if available, or by boiling, or by exposure to direct sunlight for a few hours. Louse bites should not be scratched, no matter how irritating, as this leaves the skin vulnerable to infection with typhus through the louse faeces. Instead, wash the skin with weak antiseptic or a strong soap.

When a tick bites into skin it embeds its head in the flesh. For this reason they should never be simply pulled off, as they may leave the head behind and this will cause an infection. Smothering their bodies in smoke, iodine, paraffin, petrol, etc will only make them vomit – thus, again, causing infection. The best way to remove a tick is to pinch the surrounding skin with tweezers, pulling the tick with the flesh. Apply thumb pressure to the small hole and it will stop bleeding and soon heal.

Hair Hair can attract lice, and is best kept short. It is easier if another member of the party does this, using any available scissors rather than a knife. During any long-term survival situation (more than one week) all members of the party should crop their hair. Do not discard any cut hair – it can be used as tinder.

Keeping the teeth clean reduces the risk of serious stomach upsets.

Fungal Problems The heat and humidity provide the perfect conditions for the development of fungal and bacterial growth on the skin. This is why personal hygiene is so important. Any skin that is constantly damp and not exposed to the air will invite an attack by some type of fungus. Athlete's foot and ringworm are types of fungal conditions. Dhobi itch is another, which attacks the groin leaving itchy brown or reddish areas. Fungal infections rarely clear up on their own, especially in such conditions, and may need application of a fungicidal cream over a period of time.

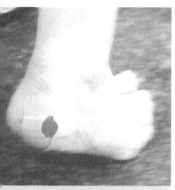

If you have to travel across country then how well you take care of your feet may be the factor which decides whether or not you survive.

Prickly heat occurs when sweat glands become blocked, causing a rash. It occurs when skin is moist and hot all the time, and is common in people not acclimatized to the tropics. Loose clothing will help to prevent this condition, and cold water poured over the skin will soothe it.

Teeth Teeth can be cleaned with an improvised toothbrush made by chewing the end of a stick to separate the fibres; use the stick only once, then discard it. Lye slurry, soap, sand and salt can be used instead of toothpaste. The inner strands of paracord or the fine fibres on the inside of tree bark can be used as dental floss. A mouthwash can be made from salt water. Painful cavities can be filled with candle wax to help relieve the pain.

Feet Feet require constant maintenance; blisters or ingrown toenails can be extremely painful,

Survival Soap

Good hygiene prevents disease and illness, and is never more important than in a survival situation. Survival soap is easy to make and will help clean wounds and wash clothes.

❶ Melt animal fat by cooking it in water while constantly stirring. Drain off the grease into a flat tray to harden (a metal wheel hub is ideal).

❷ Take a clean sock or shirt sleeve and fill it with cold, crushed ash from the fire; soak the whole sock in water, and hang it up so that the water and charcoal drip out – this liquid is potash or lye.

❸ Remelt the grease and add the lye, mixing two parts grease to one part lye. Boil the mixture until it thickens to the consistency of porridge, then allow it to cool.

❹ It can be used in its liquid form, but is best left to go solid and cut into blocks.

Diarrhoea

Charcoal powder can be purchased and added to your survival medical pack, or produced in the field when required. Small amounts can be taken dry, but it is best administered mixed with water into a slurry.

As with most foreign travellers, survivors will suffer from bouts of continuous diarrhoea. Although unpleasant they pose no threat to life, and the disorder is usually self-limiting. In the case of the survivor diarrhoea will normally develop as a result of consuming contaminated food or water, although malaria, cholera and salmonella produce similar symptoms.

Diarrhoea is detected when the number of daily bowel movements increases by a factor of two or more, the stools being soft and watery.

A small amount of charcoal slurry will settle the stomach, as charcoal absorbs toxins from the gut. Take charcoal from a cold fire, grind it to a powder and mix it with water. The thickness of the slurry is determined by its usage; for diarrhoea a light mix is required, about 10 grammes (0.35 ounce) of charcoal to a cup of water; for stomach poisoning the mixture should be 50 grammes per cup.

and may prevent a survivor from walking. Foot blisters are usually caused by ill-fitting boots, poor quality socks or loose laces, combined with long periods of walking over rough, uneven ground.

Stop and treat small blisters immediately by covering them with surgical tape. A severe blister is often filled with fluid, and can be made more comfortable if the fluid is drained. Large blisters which look as if they are about to burst should be punctured with a sterilized needle and thread. Run the needle through the blister from side to side, then clip off the thread leaving a short length hanging out each side of the blister. This will ensure that the fluid drains without creating a large break in the skin. The surrounding area must be kept thoroughly clean and dry.

Ingrown toenails should be treated as soon as possible. The best method is to shave the top centre of the nail with a razor blade. Skim the middle third of the nail, shaving from the bed towards the nail tip. Place a thin piece of plastic under the nail to prevent accidentally cutting the toe. When the nail is thin enough it will buckle into a ridge and relieve the outward pressure. Removing the nail altogether should be avoided, as this will require a dressing and prevent the patient from walking.

Wet Water Immersion Wet water immersion foot is rather like trench foot but occurs in warmer climates. Feet that are constantly immersed in water, such as when repeatedly crossing streams and swamps, will be susceptible to this condition. The soles of the feet become white, wrinkled and sore, and walking becomes very painful. Chafing of the skin can occur when it is constantly exposed to soaking wet clothing; the crotch area, in particular, becomes sore and red through the rubbing of the wet material. Chafing and water immersion foot are helped by allowing the skin to dry out and heal.

Rashes Avoid scratching any rash. Dry rashes should be kept damp and wet rashes kept dry. A small rendering of boiled animal fat and crushed charcoal rubbed into a dry rash will help prevent the skin cracking and promote healing. Fungal infections are best exposed to direct sunlight, and kept dry. Skin rashes that become infected should be treated as open wounds and dressed accordingly.

Poisoning If you are forced to live on wild plants your chances of being poisoned are greatly increased. The danger can be averted by eating only those plants or fungi which are easily recognizable. If poisoning is suspected, the patient must be made to vomit. A glass of water mixed with salt followed rapidly by gagging should produce the desired result. Use you fingers or a smooth cold instrument such as a spoon handle to stimulate the throat. After vomiting give the patient charcoal slurry, to help absorb any remaining poison.

Pain Relief Smoking the below-ground root of the Strangling Fig has the effect of relieving pain. The roots, about the size of cigars, should be pulled out of the ground, washed and left to dry. The roots contain very large water veins which become hollow

Foot Care Overnight

• At night remove footwear and socks, to give your feet a chance to breathe – but remember to protect them from biting insects if you do not have a mosquito net.

• Place your footwear upside down on two metre-long sticks pushed into the ground underneath your shelter or hammock. This will help them dry, foil some of the intrusive wildlife, and keep them where they can easily be grabbed during the night.

• Dry socks by hanging them from your hammock strings.

The dried root of the Strangling fig can be smoked to provide effective pain relief.

during the drying process; it is this that allows it to be smoked. Cut the root into lengths of about 10cm (4ins), and smoke the whole root to gain any benefit.

Author's note: Although I have never smoked and always condemn those who take illicit drugs, in the interests of survival I have tried this remedy. The experiment was a little inappropriate, as I had no pain to relieve; but there were no visible side effects. The Iban trackers of South-East Asia, who regularly smoke the root, seem remarkably healthy.

Treatment with Maggots

There has been much speculation about the use of maggots in wound treatment. They do have a value; however, they can be a double-edged weapon, and their use must be carefully monitored. They should only be applied when antibiotics are not available. Despite the hazards involved, maggot therapy should be considered if a wound becomes severely infected and ordinary debridement of rotting tissue is impossible.

1 Remove any bandages to expose the wound to flies, which will deposit their eggs on the rotting flesh.
 Warning: The flies are also likely to introduce bacteria into a wound, causing additional complications. Limit the number of flies accessing the wound – one exposure should ensure enough maggots. Live or hatching maggots will naturally find their way into the wound, at which stage the wound should be covered with a clean dressing.

2 The dressing should be removed daily to check for maggots; if none are found within three days expose the wound to the flies once more. If there are too many maggots, remove the surplus with a sterilized instrument, leaving no more than a hundred in the wound.

3 Monitor maggot activity very closely each day. The maggots produce a frothy red fluid which must be sponged away with

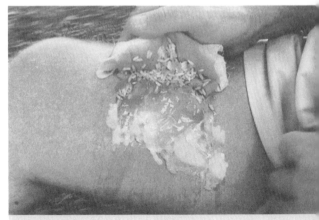

Maggot activity should be monitored closely.

a sterile cloth in order to keep track of the maggots' progress. The time taken by the maggots to clean the putrefying tissue from the wound will depend on several factors: the nature and depth of the wound, the number of maggots present, and the type of fly which laid the eggs.

4 Many people believe that maggots will only feed on dead tissue, but this is not the case – they also eat living tissue if nothing else is available. Maggots eat at an alarming rate, so the wound should be checked on a regular basis or whenever the patient feels any sharp increase in pain; this, and any fresh blood flow, are good indicators that the maggots have eaten all the dead tissue and have started to invade healthy flesh.

5 At this stage all the maggots should be removed by flushing the wound with sterile water or fresh urine; it should then be carefully sponged dry. The wound should be left open, and checked every few hours to ensure that it is completely free of maggots. Once all of the maggots are removed, bandage the wound and treat as normal.

Malaria

The parasites which cause malaria are transmitted by the bite of female anopheline mosquitoes. The parasites migrate to the victim's liver, where they multiply. After nine to 16 days they return to the bloodstream, breaking down the red cells and causing anaemia. At this point the outward symptoms appear: fever and chills, headaches, and joint pain. These symptoms may be mistaken for flu, food poisoning, or even jet lag. In danger regions, always assume that fever means malaria unless you have good reason to think otherwise. Untreated, plasmodium vivax can lie dormant in the liver for years, causing recurrent bouts of illness.

Protecting Yourself Against Malaria

- Before travelling to danger areas, check which antimalarial drugs are recommended on the web site of the Centers for Disease Control at www.cdc.gov/travel.

- Take the drugs regularly throughout your trip.

- Follow all the advice in this book on avoiding mosquito bites: use repellents, smoke, etc; cover your skin; avoid stagnant water particularly after dark.

- Most antimalarial drugs should also be taken for four weeks after you leave a danger area.

- If symptoms strike after you leave the area – fatigue, weakness, light-headedness, fever, chills, nausea – get immediate medical help. Tell your doctor where you have been travelling.

Rabies

Assume any bite suffered from a warm-blooded animal in the Third World is dangerous, even if the animal had no obvious symptoms (classically, drooling at the mouth coupled with agitated body movement and noises, and unprovoked aggression or over-friendliness).

In humans the symptoms are fever, headache, sore throat, nausea, loss of appetite, followed by pain or numbness at the infection site, skin sensitivity to temperature changes, depression and insomnia. As the virus attacks the central nervous system extreme pain is suffered when swallowing, leading to 'foaming at the mouth' through inability to swallow saliva, and terror at the sight of water (hydrophobia). Dementia or paralysis follow, and sometimes coma, always leading to death.

The great danger lies in the long incubation period between infection and the appearance of symptoms – this can vary from five days to more than a year in some cases, the average being two months. Once symptoms appear in humans there is no cure.

Treatment
Pre-exposure vaccines are available, but immunization must be planned well ahead. A course of three injections is given over 28 days. Its effect may also be impaired if you are taking antimalarial drugs. Some individuals may be allergic to compounds in one or other of the three types of vaccine (RVA, HDCV and PCEC). For all these reasons, get qualified medical advice in good time.

If caught soon enough, rabies is curable during the incubation period, so treatment should always be sought even if some time has passed since the bite. If in doubt whether an animal which has bitten you is rabid, there may be some circumstances in which you can capture it and watch it to see if it develops (further) symptoms; handle it with extreme care to avoid further infection.

A rabid animal will normally become obviously sick within about three days of inflicting an infectious bite. If you are far from any realistic hope of post-exposure treatment, it will at least reassure you if the animal does not develop symptoms. If treatment may become available, cut off the animal's head and pack it carefully – its salivary glands can be tested when you reach help.

Wildlife

Mosquitoes Any insect bite has the potential to introduce infection, but tropical mosquitoes are the carriers of several dangerous diseases and parasites which can prove fatal. Malaria, filariasis, yellow fever and dengue fever as well as various forms of encephalitis are all carried by the mosquito. Do everything within your power to protect yourself from their bites.

- Mosquitoes breed in stagnant or sluggish water and on swampy ground. Avoid making your camp near any of these; seek higher ground where possible.

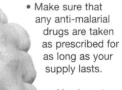

- Make sure that any anti-malarial drugs are taken as prescribed for as long as your supply lasts.

- Use insect repellent especially at night.

- Cover exposed skin wherever possible, with mosquito netting, parachute material, handkerchiefs, or improvise. Tuck trouser legs into socks and sleeves into gloves.

- Mud smeared over exposed skin may help deter mosquitoes.

- Slow-burning, smoky fires will drive insects away.

Snakes The fear of snakes is common among survivors, yet they do not present too great a risk. Of all the thousands of species less than 10% are dangerous, and generally they will not attack unless disturbed, hungry or provoked. Most snakes are very shy and will do their best to avoid you, attacking only in self-defence. Most snakes move quite slowly, and any healthy human should be able to outrun them. The one notable exception is the Black Mamba of Africa, which is extremely aggressive, deadly, and can reach speeds of up to 48kph (30 miles per hour). If you do find yourself in a snake-infested area, take a few precautions:

Remedies

• The bulbs of wild garlic can be crushed and used directly on a wound, or may be boiled to extract the oils and applied as an antiseptic.

• A handful of salt added to a litre of boiling water and allowed to cool will produce a solution that will kill bacteria.

• Sphagnum moss is a natural source of iodine, and makes a useful dressing. It is found in bogs all over the world.

• Remove the paper and tip from a cigarette and chew the tobacco until your mouth has produced enough saliva to allow swallowing. The ingested nicotine will kill most stomach worms. Repeat the process for several days until the infestation has stopped.

• The addition of hot peppers to your food diet will create a parasite-free digestive system.

- Use a stick to turn over logs and probe undergrowth.

- Watch where you walk – some snakes are dopey when basking in the sun, shedding their skin, or digesting a meal, and may not hear you coming.

- Most snakes hunt at night and prefer dark areas to rest in. When walking at night, make noise/vibration to warn them.

- If you come across a snake, stay calm and still, and then back away slowly. Avoid sudden, threatening movements. In most cases the snake will slither away.

- Never tease, pick up, or corner a snake unless you intend it to be your next meal.

Treating snakebite Bites, whether from a venomous or non-venomous snake, can be very frightening. If you cannot firmly identify the snake as a non-venomous variety, treat the bite as if it were poisonous. The important thing is not to panic. More people die as a result of panic than from the actual bite, as the increased heart rate only adds to the effects of the venom.

LIFESAVER
Snakebite: Immediate Action
- Remember that only a small minority of snakes are venomous.
- Try to stay calm and reassuring at all times.
- Position the casualty so that the bite is lower than the heart.
- Wash the area of the bite.
- Do not try to stop the bite bleeding.
- Apply a tourniquet above the bite – i.e. between the bite and the heart.
- Immobilize the limb.
- If possible, immerse the limb in cold water.

If a person is bitten the treatment must have two aims: to reduce the amount of venom available to enter the body, and to prevent the venom which has entered the wound from spreading through-out the system too fast. This will give the body a chance to deal with it at a reduced rate. Resassure the casualty; make them rest. Find the puncture wounds, and position the patient so that the bite is lower than their heart. Wash the surrounding skin with water to prevent more venom entering the wound.

Despite what you have seen in the movies, never cut a snakebite, as this may introduce more venom into the bloodstream. Never try to suck the poison out; it may be absorbed through the lining of your mouth. Instead, use a bandage as a light tourniquet; put it on above the bite, and wrap it downwards towards the puncture marks. Make it tight enough to stop the return of venous blood, but not so tight that it restricts the flow of arterial blood.

The wound will probably bleed, but this is potentially beneficial as the blood will carry away some of the venom. Put a splint on the limb to lessen any unnecessary muscle movement, which will act as a pump for the blood in the veins. If you have the means, immerse the wound area in cold water; this will further slow the blood flow back to the heart. Continual reassurance of the casualty is vital; do not let him believe he is going to die, or this may lead to shock. After 15 minutes, if the bite area is not painful or swollen and the casualty has no headache or dryness of the mouth, it is safe to assume that the bite was not poisonous.

Ants Tropical ants, both large and small, can inflict severe bites, and if disturbed will attack in numbers. The bite of the fire ant, in particular, is noted for causing excruciating pain. Unless you need them as a source of food you are best advised to leave ants alone. Both termites and ants move in continuous columns vary-ing from a few centimetres to over a metre wide. If undisturbed they will pass by, and any isolated insects which crawl onto you can usually be brushed off without biting. Ants can be a problem for injured people who are unable to move, so always lay them in a place free of ants.

Leeches Leeches are found in tropical and sub-tropical lowland forest, especially after rain. They look like fat black worms, and live by sucking the blood of any warm-blooded creature. Their bites, although not especially dangerous, cause discomfort and loss of blood and open the way for infection. They wait for their prey in swamps and slow-moving water, or attach themselves to vegetation. Check every few minutes to see that leeches have not attached themselves to you or your companions – their saliva has anaesthetic properties so you may not feel their bite. (If alone, you can use a heliograph mirror to check your back.)

As long as the leech has not yet latched on you may flick it off; however, if it has already bitten do not try to pull it off – if you do, its jaws may remain in the wound and cause infection. The best way to remove a leech is to put salt or ash on it, or approach it with a glowing ember (such as the tip of a cigarette). You can also make up a nicotine solution from left-over unburned tobacco wrapped in a piece of cloth; when this is moistened and squeezed onto the leech it will loosen its hold. Leeches will fall off naturally when they have had their fill.

Leech bites can be cleaned by gently squeezing the area, allowing the blood to carry away anything nasty. Despite the anticoagulant properties in the leech's saliva the bleeding will stop after a while and a clot will form. Leave this in place for as long as possible to protect the wound from infection.

Use your hands to scoop up water to drink – never put your face into leech-infested water. If a small leech gets into your mouth, throat or nostrils it could be very unpleasant and even dangerous. If it does happen, however, gargle or sniff up a concentrated salt/water solution to dislodge them.

Flukes and Hookworms
These tropical parasites can easily enter the body by piercing the skin. They remain in the bloodstream, and are the cause of serious, debilitating diseases such as bilharzia (schistosomiasis). They are found

Leeches are annoying but they cause little harm.

Wild animals

Attack by wild animals is less of a risk than is commonly thought. Most animals – even predators – are shy of man and are not often seen. Unless cornered, startled at close quarters, or injured, they will seek to escape. Females with young are also likely to be protective and will attack anything perceived as a threat to their young. Large cats such as lion, tiger, leopard and jaguar will sometimes attack a human who appears vulnerable or weak, or if that animal has had a taste of human flesh before.

Far more dangerous are animals such as the hippopotamus, which accounts for more human deaths in Africa than any other animal; or even the placid-looking water buffalo of South-East Asia. These are not meat-eaters, but can be extremely aggressive when roused or when feeling threatened. Elephants, too, should not be provoked, and should be given way to at all times. Unless there is reason not to, make as much noise as possible when passing through the jungle; this will warn many animals, who will simply avoid you. The risk from animals and insects will depend largely on which continent you are on.

only in fresh water that is slow-moving or polluted, so be aware of this if you have to wade such water. Always check before bathing, and do not drink water from such sources without thorough boiling.

Bees, Wasps and Hornets
These stinging insects can all be very aggressive in defence of their nests. Nests can generally be identified as brown-coloured oval or oblong masses attached to tree-trunks or branches at heights between 3m and 10m (10-30ft) above ground. If you are only a few metres away when a swarm becomes disturbed, stay very still, preferably sitting down, for at least five minutes. Once the worst threat has passed, crawl away slowly and carefully. Should a swarm attack you there are two ways to defend yourself. One is to run through the thickest and bushiest foliage you can find – the foliage will spring back after your passage and will beat back and confuse the insects. The other method is to immerse yourself in water completely, and stay under for as long as you can.

Spiders Of the many species of spider in the world only a tiny minority are deadly poisonous, such as the Funnelweb of Australia. The Black Widow and its tropical relatives also present a threat to life, but only in extreme cases is their bite fatal; it is, however, very painful, dangerous and disabling. Apart from the Funnelweb, which is grey or dark brown, most of the other spiders to avoid are dark in colour and patterned with red, white or yellow spots. If in doubt, assume that it is poisonous. Anyone bitten by a poisonous spider during a survival situation should be treated as if for snakebite.

Rice Borer Moth This insect is found in the lowland jungles of South-East Asia. It is small, plain-coloured, and has a pair of tiny black spots on the wings. At night

Military personnel often spend many hours wading through swamp areas while on exercise in tropical regions, so are all too aware of the dangers that water harbours.

it is attracted to lights in large numbers. If one lands on you it should never be roughly brushed off: it has small barbed hairs on its body, which may enter the skin and cause a painful sore which takes weeks to heal.

Water Hazards Crocodiles and caymans are flesh-eaters but will rarely attack humans; nevertheless, they are best avoided. Tropical waters provide far more dangerous but less obvious creatures than crocodiles. Piranhas inhabit the rivers of the South American rain forest, and may attack a human; they are most dangerous in shallow waters during the dry season. Electric eels can reach 2m (6.5ft) long; the larger specimens can generate 500 volts of electricity and are extremely dangerous. Catfish and poisonous water snakes can all be found in tropical rivers and may present a danger to the unwary. One to beware of is the Amazonian Candiru, a minute type of catfish. It is reported that this will actually swim up the urethra of a person urinating in the water; once inside its back-folded dorsal fin acts as a barb, causing extreme pain and a very serious medical problem. You should remember this image when wading.

The People of the Forest

Jungle survivors have a good chance of contact with people living deep in the rain forest, who in most cases have no communication with the outside world. If the local tribespeople are friendly towards you then you may well be offered hospitality. Traditionally jungle tribes are shy, but if approached in a quiet manner will accept a stranger into their midst. Unfortunately, in recent decades some innocent tribes have been exploited during political uprisings, and their jungle villages have been used as training camps for guerrilla warfare. Others have been provoked into warlike attitudes by the expansion of commercial logging and farming interests into their forests. A knowledge of the current political system governing the region should indicate to any survivor what attitude can be expected when contact is made with the forest people.

Having assessed your situation, you must judge the urgency of your need for shelter. If you decide to stay where you are rather than walking out, then it is worth planning and constructing a more substantial shelter.

Raise your sleeping place well above the jungle floor.

Given the amount of materials readily available, with a little initiative you should be able to quickly build a shelter which will considerably increase your level of comfort. Shelters are necessary to give protection from the heavy downpours, and for getting a good night's sleep. As well as following the general principles of shelter-building, in the jungle your camp should be sited on high ground away from swamps and dry riverbeds alike.

The term *basha*, meaning a jungle dwelling, has been adopted by the SAS for any form of sleeping place. Follow the SAS example, and always check the site of your basha first for snakes, ants, ticks, standing dead trees, and anything else that might crawl into or fall onto it. It is wise to check the surroundings for any large, well-worn game trails, taking note of the size of any pawprints.

If you can, raise the whole shelter off the ground; if this is not possible, then at least make sure that your sleeping area is raised – do not sleep directly on the jungle floor, which is alive with biting insects and home to other kinds of wildlife with which you will not enjoy sharing your bed. Consider all possible shelter options, from aircraft wreckage or parachute material to building a tree house (which is easier than it sounds).

Large aircraft crash-landing in jungle are usually completely destroyed on impact (the consolation is that they leave a clearly visible trail). Occasionally large sections remain intact enough for use as shelter by survivors, however. Make sure the airframe is firmly settled and will not move further.

Sleeping in the Jungle

• Do not camp close to swamps, dry water-courses, or large animal trails.

• Camp on high ground where possible.

• Check the ground and vegetation for snakes and insects.

• Check above the site for rotten branches, and wildlife.

• Raise your sleeping place well above the jungle floor.

With a good hammock and poncho, jungle survival is easy.

Remove seats to make space for sleeping, or for any injured; and make the fuselage as insect-proof as possible.

Hammock If you arrived in the jungle on foot you should be carrying a purpose-made hammock or pole bed. In the unlikely event that you parachuted in, then you have the means to make one. Improvise using some form of fabric and cordage from a crash or abandoned vehicle; hammocks made from vines tend to break, however, so it is best not to attempt this method.

Jungle Materials

The forest abounds with building materials. Young saplings between 8cm and 10cm in diameter (3ins-4ins) make excellent framework poles. All species of bamboo are found in tropical or semi-tropical regions where there

The jungle is full of large, broad leafed plants which are ideal for shelter construction.

is sufficient rainfall. They are tall, fast-growing members of the grass family. The mature woody stems – which in some species can grow to a height of 25m (80ft) – are jointed and hollow. Bamboo is extremely useful as a building material or for making utensils. Beware when harvesting bamboo, however – the tight-packed clumps of stems are often under great tension and may suddenly shatter or whiplash as you cut them. This can cause serious injury, as bamboo splinters are very sharp; and note also that the bases of bamboo shoots are covered with fine, stinging hairs which may cause skin irritation. Vines make good tying material and can be pulled down from trees whenever required (but be careful to look upwards, and make sure that it is only a vine that you are pulling down).

Sheeting Plastic or canvas sheeting of any kind can be used in the construction of a shelter – ground sheets, parachutes, plastic sacks, jute sacking, tarpaulins and blankets can all be used in some way. Wreckage from vehicles and aircraft can all be put to use providing excellent weatherproof shelter.

Shelter Construction

Pole Bed My advice is to build your pole bed first and then construct a shelter around it – insertion may be tricky if you work the other way round. The bed itself can be made of bamboo or other small branches lashed together to form a frame – either a freestanding A-frame or one integrated into the hut itself. A simple A-frame is made by lashing two poles 2m (6.5ft) long together at the top and splaying the legs apart. With two frames, each resting against one of a pair of closely spaced trees, the bed frame can be dropped over the top; the height of the bed is adjusted by opening or closing the A-frame legs. When it is satisfactory, tie the whole thing together with cords or string vine, and fit a poncho as overhead cover.

Attap Basha Many broad-leaved plants grow in the jungle, but attap is the most commonly used in shelter construction. The branch has V-shaped leaves growing down a central stem; although sharp, with careful handling these can be used to thatch shelter roofs and walls. Neatly interwoven on frames of bamboo

A tree hut offers protection from beasts and insects, and is cooler at night.

Bamboo

• You can use lengths of bamboo to make any kind of shelter, platform, bed, raft, etc.

• You can use short sections cut close to the joints to make cups, water carriers, food containers and cookers.

• You can use split bamboo to make eating utensils, skewers, etc.

• When harvesting bamboo in thickets, check first that your cut will not suddenly release a stem under tension.

• Avoid cuts and punctures from sharp bamboo splinters.

Simple lean-to with bamboo floor.

or saplings, attap is a versatile building material used by villagers throughout South-East Asia. It is often possible to find four closely-growing trees to form the basis of your shelter without felling; these will give the firmest possible support.

Ground Hut This is the attap variation on the North American tepee, using a framework of poles angled together and tied at the apex. This can be covered with parachute material or attap branches. Cross-poles should be incorporated to form a raised platform inside, for seating by day and a bed by night.

Tree Hut Tree huts are common in South America and Asia, and are built to protect the inhabitants from flooding and dangerous wildlife, or simply to give them a different perspective from that of the forest floor. One tribe in Borneo build their homes on the roof of the jungle canopy, interconnected by swaying ladders and platforms of vine. For survival purposes there is no need to build more than 5m (16ft) above the ground. Cut your frame poles on the ground, leaving them long enough for an individual to reach down and pull them up to the construction platform. Once a level base platform has been completed, build an attap hut on it as normal. A long, thick log inclined against the tree and with steps cut into it will form a simple ladder.

Swamp Bed If you are forced to camp in a swamp or on habitually wet ground and have no hammock, you must construct a sleeping place above ground. Take into account the location – e.g., if near a coastline you must allow for tidal changes. Try to use the surrounding vegetation. For instance, in many swamps the trees can be very large, with fin-like buttresses at the base; by cutting into these fins it is simple to form a platform big enough to sleep on. Another method is to link two nearby trees by securing a pole at the same height on either side, and criss-crossing these with shorter lengths to form a platform. If sheeting is available, place another pole 1m (3.25ft) above the sleeping platform to sup-

A Jungle tepee.

port overhead cover. Bamboo is ideal for this sort of work, but any strong pole about 10cm (4ins) in diameter will suffice.

Finding and Making Cord All survival kits should include a length of parachute cord; but

A basic pole bed is designed to get you off the jungle floor.

Cordage is a central necessity for most survival tasks, from mending clothing and shoes to constructing packs, traps, tools and shelters.

many survival situations will also reveal several ways in which cordage can be produced. Parachutes contain at least 100 metres of excellent cord, from which the thin, inner strands can be extracted. Electric cable and control lines are also a valuable source of material for making lashings. Many military combat belts are made up of a series of strong cotton strips, and the material can be unravelled to make cordage.

Natural materials such as animal sinew and gut make ideal cordage. These are best dried and separated into the required thickness. Wetting them before use will allow for easy manipulation, while their hardening when dry will hold any knot firmly in place. The bark of some trees can be used, either in its natural state or split into stringy fibres.

Knots Knots and lashing are best kept simple, and require little or no explanation. If the survivor is unable to tie a secure knot then his chances of survival are fairly limited. It should also be remembered that any knot or lashing is only as strong as the material used in its construction. Knots and ropes are used in climbing, shelter construction, and a whole host of survival applications. A knot can join one or more ropes together, either permanently or for quick release depending on the requirement.

Utensils

Billy Can A 'billy can' is perhaps the most important utensil for the survivor – a container which can be used to collect water and plants, and also for cooking. Any metal container, such as a commercial-sized baked bean can, makes one of the best, with a wire handle attached for carrying.

Bamboo Bamboo has many uses, as jungle tribes have discovered over millennia. It can be used to make rafts, construct homes, build traps, and – because of its sectional nature – to make carrying, cooking and eating utensils.

Birch Bark Containers Containers made from birch or cherry bark will not burn through when heated over a moderate fire, provided that you fill them with sufficient water. Cooking on a fire of

Improvised Tools

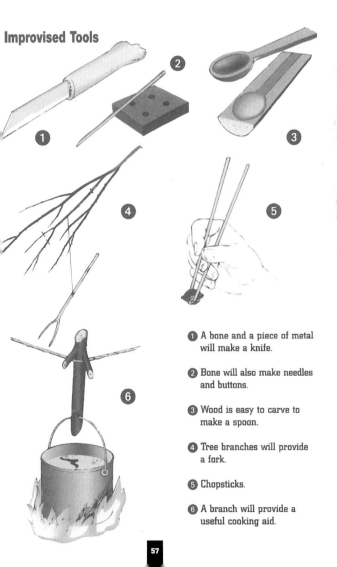

❶ A bone and a piece of metal will make a knife.

❷ Bone will also make needles and buttons.

❸ Wood is easy to carve to make a spoon.

❹ Tree branches will provide a fork.

❺ Chopsticks.

❻ A branch will provide a useful cooking aid.

Know Your Knots

The Reef Knot

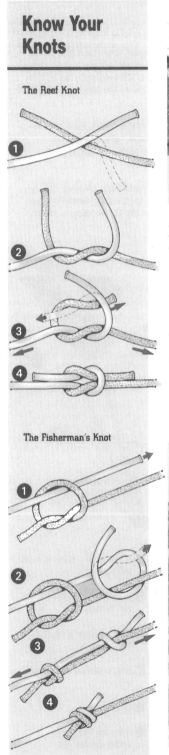

Birch bark makes waterproof containers which can even be used for cooking.

glowing embers produces a better meal and extends the life of the bark container. Shoes can also be produced from such bark.

Eating Utensils If you have a penknife, you can construct a simple spoon from a flat piece of wood. Mugs can be made from a section of bamboo, a carved-out piece of wood, or a folded piece of birch bark. Many naturally occurring items can be fashioned into simple but adequate eating and drinking utensils with a little ingenuity and experimentation.

After a knife, a billy can of some sort is probably the most constantly used item in a survival situation.

The Fisherman's Knot

Tyres In survival situations brought about by vehicle accidents the survivor may well have access to tyres, which are a valuable resource. Tyre rubber can be cut up to make shoes and belts, and the reinforcing wire can be stripped out for traps and snares. Burning tyres create large amounts of black, acrid smoke which is excellent for signalling; for this reason they should normally be kept for use in rescue beacons rather than used for everyday fuel. (However, they could be used in an emergency when fire was a question of life or death, e.g. after falling through ice.) A tyre set in the middle of a three-pole frame makes a comfortable toilet seat; its value to morale should not be dismissed, particularly where survivors are suffering from digestive disorders.

Maintain a routine in a long term survival situation.

Long Term Camp Routine

Long term camp routine implies that you will be permanently in one place for an unspecified length of time. To alleviate boredom and maintain a sense of personal discipline and hygiene certain routines need to be established, both for the individual and any group of survivors.

• In order to improve your existence you must become a scavenger and improviser. The world is full of rubbish discarded by thousands of travellers. From the inner forests of the Amazonian basin to the peaks of the Himalayas, and even in the vast wilderness of the polar regions, you will find the pollution of discarded man-made artefacts. To the survivor all this rubbish has a value.

• Long term camp sites can easily become contaminated with urine and faeces, and a strict routine should be organized as soon as your shelter and fire are finished. Dig toilet holes until a more permanent structure can be built. Collect your water upstream, and locate your washing point at least 50 metres downstream.

• Organize your day around the priorities of rescue, health and survival.

• Check your signal fires and markers, and always have your heliograph and signal flare with you at all times – Sod's Law dictates that you will be out of camp when the search aircraft flies over.

• Establish a disciplined pattern of working to prevent boredom and the consequent drop in morale. Get up early; make a warm drink; check your traps; collect firewood; make something useful.

Fire in the jungle is mainly important for cooking and water purification. Although much of the jungle is damp, there is an abundance of rotting vegetation and no shortage of fuel.

Practical and Morale Value

Fire has many obvious practical uses. Heat sustains wellbeing and life itself in cold or wet environments. Fire can be used to cook food, to dry clothes, to purify water and sterilize medical instruments, and to signal your rescuers. Waterborne diseases are one of the greatest dangers to survival (see Water section), but boiling will kill most harmful organisms. Hot drinks provide a vital source of body heat. Cooking food not only makes it more palatable, but also destroys many harmful organisms in animal products and neutralizes the toxins found in many plants.

Fire also plays an important psychological role in survival. Being able to build a fire proves to the survivor that he can control at least some elements of his situation and provide himself with the comfort of warmth and light. He will feel that he has achieved something positive by bringing back a hint of normality to his life.

To make a successful fire you need three elements: heat, oxygen, and fuel. If any element is missing your fire will not burn. However, before you even start to build any fire consider the following questions:

- Does the time you intend staying in your present location justify a fire?
- Do you really need a fire?
- Is there enough fuel nearby to sustain a fire?
- Are you in an area where fire could easily spread out of control?

Fire-starting Materials

Building a fire calls for an understanding of the dynamics involved. When any fuel is burned, part of the heat from that combustion will go on to ignite the next piece of fuel. The hotter a fire the better it will burn.

You do not need a great deal of heat for the initial ignition – a match is usually enough. However, because the first heat source is so small and lasts for only a short time, the material you apply it to must ignite very easily. This material we call tinder.

'Feathered' sticks make excellent kindling.

Tinder Whatever its source, tinder must fulfil certain criteria if it is to ignite readily. It must be bone dry and small in size, and must readily accept flame. Ideally the tinder should burn quickly, producing maximum heat. Included with tinder are certain combustible fuels; these may be in liquid, gel or solid form and are mostly man-made. Using ammunition or flares to start a fire can only be justified

Tinder Sources

Manmade:
• Petrol, paraffin or aviation fuel.

• Oil (needs heating first).

• Cooker gel or solid fuel blocks.

• Propellant explosive from ammunition (obtainable, with care, by prying bullet/shot out of cartridge case).

• Pyrotechnics – flares, etc.

• Tampons (check with any female survivors).

• Cotton wool (check any injured survivors for useful dressings).

• Lint from twine, canvas, bandages, etc.

• Scorched or charred cloth, especially linen.

• Charred rope.

• Some photographic film.

Natural:
• Decayed or powdered dry wood and pulverized bark.

• Catface (the resinous scab found on damaged ever-green trees).

• Coconut palm frond (the fabric-like material at the base needs to be sun-dried).

• Dried Arctic cotton grass or moss.

• Termite nest material.

• Birds', rats' or mice nests.

A successful fire is best started when it is shielded from the elements with fuel added gradually.

after carefully weighing the value of saving them for their original purpose. (See box for a list of tinder sources.)

Kindling Kindling consists of material larger in size than tinder but smaller than the main fuel to be used on the fire. Ideal candidates for kindling are small dry twigs, or shavings made from dry sticks, a process known as 'feathering'. Once your kindling takes hold, the fire should burn long enough to deal with small logs, i.e. the main fuel. Starter wood for fires needs to be dead and dry.

Fuel Your fuel should be graded and stacked ready before you start, with dry, dead material separated from green wood. A hot fire will be able to cope with green logs, as the flames will boil the sap away and dry the wood before it burns. However, green logs will not catch on a fire that is not well-established and hot.

Heaping the fuel on too quickly will kill a fire. Build your fire with care, adding more fuel only when the previous fuel is burning well. Do not stifle the fire by depriving it of the oxygen it needs – make sure it is well ventilated.

In principle, the harder the wood the longer it will burn. Try to use fuel that is close at hand, still standing, and does not require

LIFESAVER

Carbon monoxide poisoning occurs when incomplete combustion of fuel takes place. In an unventilated enclosure it can quickly become lethal, especially when refined fuel such as petrol or aviation fuel is used. The following rules should be taken into account:

❶ Always ensure good ventilation if using an oil burning stove.
❷ Check any manufactured stove, and maintain its clean running.
❸ Always turn off or extinguish any petrol or aviation type stove before going to sleep.

chopping. If a log is too large, drag it into the fire and let it burn through the middle. Fuel taken from the forest floor will burn if stacked above ground for a few days, or placed around the edge of the fire to dry.

• Birch burns best.

• Both oak and ash burn well and give off good heat.

• Fruit trees such as apple and cherry give off scented smoke.

• Lighter woods such as larch and pine will spit sparks.

• For a concealed fire, burn elm.

Setting a Fire

The site for your fire must be chosen carefully, especially if you plan on building a shelter or if a strong wind is blowing. The heat should provide warmth for your shelter, but in such a way that the smoke does not envelope you (though if biting insects are a problem, a little smoke will help drive them off).

Constructing a windshield will prevent the wind from blowing out the first fledgling flames. It will also cut down the amount of fuel consumed, and reflect extra heat into your shelter.

The ground below your fire should be dry and clear of vegetation to stop the fire spreading.

If stones are available, build a circle around the fire once it is well alight. This reduces any danger of the fire setting your shelter or surroundings alight; it also defines the fire's size and fuel consumption.

If not maintained, such as overnight or while away hunting, a fire may well go out. Most fires can be relit by placing a small amount of tinder on top of the old embers and blowing. The earth below an old fire site will stay warm for many hours; this will help generate a new fire quickly.

Fire on Wet Ground To build a fire in swampy conditions the base must be raised above the water. In some cases this may mean building a platform of old logs or stones on which the fire will rest. In extreme conditions a platform can be constructed

Heat Reflection

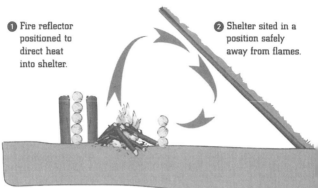

1 Fire reflector positioned to direct heat into shelter.

2 Shelter sited in a position safely away from flames.

several feet above the ground. One tribe who live in tree houses amid the jungle canopy of Malaysia cook with open fires using a base of stones and baked earth spread over the bamboo floor.

Fire in the Wind If the weather is extremely windy, a fire-shield will do little to stop the flames from getting out of control or being extinguished. In such conditions the only answer is to build your fire well below ground level, by either digging a trench or finding a natural ground hollow.

Ember Pit No matter what type of open fire you make, they are all difficult to cook on. Either they will burn the meat, or you will get burnt trying to rescue your supper. Metal cans become hot and the danger of scalding is inevitable. Rather than struggling with an open fire it is a good idea to make a small ember pit for cooking. This is simply a matter of cutting out a section of turf 20cm long by 10cm wide and 10cm deep (8ins x 4ins x 4ins). Once your fire is well established, use a stick to rake glowing embers into your pit. These will supply a manageable source of heat for cooking. As the embers die down or you require more heat, simply rake in more embers.

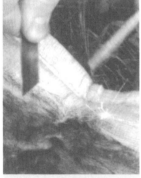

Sparks from a flint-and-steel were man's only portable fire-starting method for many centuries.

Lighting your Fire

The initial heat source for a fire can be produced in any number of ways. Matches or lighters provide the easiest option, but these will not last forever. Tinder can also be ignited by sparks from a flint- and-steel set, or from an electrical source such as a car battery. Heat from the sun can also be concentrated and focused by a magnifying glass or a parabolic reflector. Most heat sources are derived from commercially produced items, but if these are not available sufficient heat to light a fire can be generated by friction.

Matches Matches are the most convenient and obvious way of initiating a flame, and it is a matter of common sense that they should be carried as a matter of course on all outdoor trips. Ordinary matches do not work when damp, however, and can be quickly extinguished if unprotected from a strong wind. This fault can be remedied by dipping each match halfway into some molten wax. To protect the outside of the box, spray it with hair lacquer. Specially made survival matches are protected by a waterproof container, and when lit they will burn for up to 12 seconds in just about any weather conditions.

Lighters In any group of people several will probably be carrying cigarette lighters. These make an excellent survival aid, but must be used wisely and economically. Once the lighter fuel is exhausted do not just throw the lighter away – its flint will go on making sparks for a long time. A new device has also appeared on the survival market recently which converts a standard lighter into a mini-blowtorch.

One Match, One Fire

You can save on matches and lighter fuel by lighting a candle with them immediately. This candle can then be used to provide a constant flame to ignite tinder, even when it is still a little damp. Like most naked flames, the candle should be protected from the wind by a shelter. Either dig a hole into the ground or build a small stone wall around it. Place the tinder over the flame, either by piling it on top of the shelter or building a small 'wigwam'. Once the tinder has ignited, remove the candle and if you no longer need it, blow out the flame and keep it for the next fire.

Burning Glass Using a burning glass will require strong sunlight, but it can prove an effective way to light a fire given the right conditions. For the glass you could use a magnifying glass, or a lens from a camera, binoculars, spectacles or a compass. An ideal size would be 5cm (2ins) or more in diameter. Sunlight focused through the glass will ignite dry tinder, although you may need to fan it lightly as it smoulders.

Batteries If you have access to a large capacity battery from a vehicle, even if the vehicle has broken down or crashed, you may be able to start a fire by electrical means. Use a thin wire to connect the negative and positive terminals on the battery; this will short circuit the battery and cause the connecting wire to spark.

If very thin wire can be found, roll this into a ball and touch either end with both terminals; this will cause a flash bulb effect.

A burning glass will light tinder if the sun is strong enough.

Car Battery Method

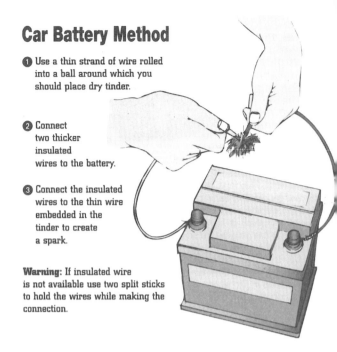

1 Use a thin strand of wire rolled into a ball around which you should place dry tinder.

2 Connect two thicker insulated wires to the battery.

3 Connect the insulated wires to the thin wire embedded in the tinder to create a spark.

Warning: If insulated wire is not available use two split sticks to hold the wires while making the connection.

Flint and Magnesium Fire Starter The specially manufactured flint is embedded into a small block of aluminium and magnesium metal which has a serrated steel striker attached. Shavings from the block can be scraped off and mixed in with any tinder. When the steel blade is struck sharply against the flint, sparks are produced which ignite the tinder. Magnesium burns in excess of 5,000° F, which is hot enough to ignite any tinder even when damp. In an emergency, scraping aluminium shavings from the frame of a crashed aircraft will produce very similar results.

Parabolic Reflector In hot, dry conditions with bright sunlight you can start a fire using a parabolic reflector (although in such conditions a fire may not be a necessity, and fuel may be scarce).

LIFESAVER
Highly Inflammable Materials

Highly inflammable materials make very effective tinder, and if near a vehicle or aircraft always check to see if any are available. Sumps, fuel tanks, lubricants, alcohol, the propellant explosive from cartridges and pyrotechnics all burn.

• **Handle them all with great care.**
• **Metal pots or implements should never be used when mixing chemicals.**
• **Many such substances will give off toxic gases when they burn.**
• **Remember that many chemicals which simply burn when loose are highly explosive when compressed or confined.**

Illustrations in some survival books depicting a hand torch reflector being used in this way are highly improbable – your best bet is to use a vehicle headlamp. Place your tinder in the bulb housing; a very effective reflector can be achieved by removing the headlamp glass and replacing it in reverse, i.e. concave. Positioning a magnifying glass on the top of the headlamp and aiming it directly at the sun will cause any tinder to ignite instantly. Water can be boiled in a similar manner.

A cane and birch torch.

Chemicals Propellant explosives can be extracted from small arms ammunition or shotgun cartridges. It is best used by sprinkling it over dry tinder and applying a spark (beware – even small concentrations of such chemicals are potentially dangerous). Simple gunpowder – 'black powder' – is not found in modern cartridges; but it is mixed from equal amounts of potassium nitrate, sulphur and charcoal. Another mixture with a high output of heat is sugar and sodium chlorate (found in some commercial weedkillers) mixed in equal parts. This will be hot enough to light a fire even from damp tinder.

Fire Paste Fire paste is highly volatile and should be kept for emergency fire-lighting only. It is basically any combustible material that is held in a suitable base: aviation fuel mixed with soap is one example. A small spark will normally ignite the paste, which will then burn for several minutes. As with any combustible material, the paste should be kept in an airtight container when not in use.

In a dire emergency the paste can be burnt purely as a fuel.

Fire Torch Cane has been used to make torches since prehistoric times. One end is split and separated in order to hold some form of burning material. Where the wind is not troublesome this can even be a candle. Birch bark rolled, split and dried also makes a good torch.

Carrying Fire

One sure method of making a quick fire especially in the cold and wet is to carry embers from your previous fire. Many early hunters transported fire around in a cow or buffalo horn.
This method can still be used simply using a perforated beer or coke can as a fire carrier. The skill remains in packing the embers from last night's fire onto a bed of dry, slow-burning material, and covering them with the same. The secret lies in the ability to maintain the correct amount of oxygen being fed to the embers. If they are wrapped too tightly, they will be starved of air, if too loose there is a danger that the embers will ignite the surrounding material.

Fire from Friction

Creating fire from friction may be the only method left to a survivor, particularly when his consumable means of fire-making, such as matches, have run out. Many primitive peoples around the world still light their fires by friction; it is presumably the oldest method of creating a flame, and certainly dates back to our prehistoric ancestors. Although methods vary from continent to continent, the following covers the basic principles of fire by friction.

Fire Plough The fire plough method involves rapidly rubbing a hardwood shaft against a softwood base. Under ideal conditions both woods should be seasoned so that the moisture content is minimal. In an emergency this can be achieved by sun-drying green wood, although it will take several days.

The baseboard should measure around 30cm by 10cm (12ins by 4ins) and have a straight central channel cut down the entire length of one side. One end of the shaft should be rounded to fit into this groove, and ground up and down the baseboard channel – adding a little sand will speed up this process. Once both the tip of the shaft and the channel have become blackened and smoke can be seen rising, the fire plough is ready for use. Kneel and place the baseboard against the left thigh. Grip the shaft with both hands and make a sharp, stabbing, ploughing action. As you build up speed small particles of wood fibres will fall to the ground. Place a small amount of dry tinder at the base of the channel ready to catch these. Once the tinder is smouldering, blow on it until you have fire.

The fire pump or piston, an ingenious invention thought to have originated in the Orient.

Fire Piston or Pump To the best of my knowledge the fire piston evolved in the Far East. It might have been invented by the Chinese or possibly the Japanese, who until recently commonly used brass cylinders of a similar type for lighting cigarettes.

The fire pump requires a body into which a 12cm (4.75in) long chamber has been reamed, similar to the barrel of a gun. A piston, with a handle to assist pressure, is inserted into the chamber. The end of the piston is cup-shaped, and into this a small amount of dry tinder is placed. The piston is then thrust rapidly into the chamber forcing the air molecules to compress, generating enough heat to convert the tinder into glowing embers.

The walls of the 'barrel' have to be straight, smooth and lightly greased. To assure an airtight fit, thread should be carefully wrapped around the piston about 2cm above the tinder cup, and this too can be lightly greased. The tinder needs to be bone dry and extremely light. The downy vest from between the layers of a banana palm stem, dried in the sunlight, is used locally in the Far East; however, during experiments I found that dried mushroom, cotton wool and even charcoal will ignite in a well-made fire pump. Making my own fire piston brought me to the conclusion that it needs to be a precision-made device. That said, a good one would last almost indefinitely.

Experience and precision play a large part in getting the fire piston to work. Ram the piston all the way down the chamber with a half-blow, half-push motion, then withdraw immediately. Keep the withdrawn piston upright so that the glowing embers don't fall out.

Bow & Drill The basis of this classic method is a flat dry board of powdery wood and a hardwood stick. The principle is to make a hole in the flat board into which the hardwood stick – i.e. the drill – will fit neatly. The baseboard can be any length, but 30cm by 10cm by a minimum of 2cm thick is ideal. Cut a V-shaped notch about 2cm wide on one edge.

The drill should be made from a length of medium-hard wood such as elm, willow, cedar, cypress, cottonwood or balsam fir. Make sure that the chosen piece is sound and dry, and that you are able to cut a straight length from it 20-25cm long and 2cm thick. Sharpen one end of the drill to a 45 degree point and the other end to a 60 degree point.

The baseboard should be placed on the ground and held in place by the toe of your boot. Place a small ball of tinder directly under the notch cut in the baseboard, dropping a little into the notch

Using a Fire Drill

• The ideal wood for the board is old deadfall, dried-out and powdery.

• The ideal drill is a sound, dry length of hardwood.

• Use your best available tinder.

• Settle yourself comfortably, so you can drill for hours if necessary.

• Don't give up – you will need hours of patience.

• Have extra tinder handy to add when you see smoke.

• Have fine, dry kindling handy to add as soon as you have blown the sparks in your tinder into flame.

• Shelter and nurture your burning kindling until you can add small, dry fuel.

Using a fire drill. It may take hours of work before the drill starts to smoke.

itself. The best tinder is dried grass mixed with small strips of cedar or birch bark. Fit the sharper end of the drill into the notch of the baseboard, and prepare to rotate it backwards and forwards. This can be done by rubbing the drill between your palms, but using a bow is better.

Make a small bow about 60cm-70cm long, and loosely string it with a length of cord or leather thong. Twist a loop in the bow-string and slip it over the drill. Pushing the bow back and forth will rotate the drill in the base-board notch. Extra pressure can be placed on the drill by using some form of cup over the upper end to hold it in place; in my experience a near-perfect expedient is a small glass jar, as used to hold fish paste.

• Do not expect to make a fire instantly; the drill will need to 'bed' itself into the baseboard, and this can only be achieved after hours of constant drilling.

• An indication that you are close to producing hot embers will come when the drill becomes charred and smoke can be seen rising from the notch. Add a little tinder to the notch, and work the drill vigorously. Embers from the notch should fall into the tinder below. Carefully move the block away and blow gently on the pile until the tinder ignites. As with any survival skill, practice makes perfect.

A star fire.

Types of Fire

Once you have a fire, you need to make sure that it is suitable for your needs. If you are alone you will only need a small fire for warmth. Small fires need less fuel to keep them going, and can be controlled more easily.

If it is snowing or raining – and if you have the means – you should consider taking your fire inside your shelter; and the best way to do this is to improvise a stove. You can also build a fire that will cook your food while you are hunting, and warm your bed on a cold night.

Star Fire This is a simple and easily controlled fire. Once the fire is established, place logs so that they can be fed inwards, increasing the flames. If less heat is required the logs can be pulled outwards. You can leave this fire for several hours while you go hunting. The flames will eventually die down, leaving the hot embers in the middle; these can be protected from wet weather by placing a large stone over the inward ends of the logs. When you return, carefully remove the warm stone and use it as a seat. To rekindle the fire simply push the logs closer together and gently fan or blow on the embers.

Lumberman's Fire A lumberman's fire is built using two long logs – the larger branches from fallen trees are ideal. The purpose is to build a normal fire between the logs until the fire has reached the point where the logs themselves will burn. If the timing is right you should be able to cook your food on the small fire in between before rolling the logs together. You will then be able to stretch out for the night along the length of the burning logs, and have a good sleep. Separating the logs in the morning and adding a few twigs will quickly rekindle the fire for breakfast.

A pyramid fire.

Fire In Swamp Areas

In wet areas build your fire on a platform well above the water surface; use poles and other vegetation, finished with a thick pad of mud or silt on top of which you can set your fire.

Fire As Insect Repellent

The smoke from a burning termite's nest will keep biting insects away. Crawling insects do not like walking over ash, and it is a good idea to spread some around your sleeping area at night.

Spontaneous Combustion

A survivor will seldom be carrying the right chemicals to produce spontaneous combustion. If, however, the right substances are available – e.g. antifreeze from a vehicle radiator, potassium permanganate from the medical kit – you can use them to start a fire.

Antifreeze will contain enough glycerine to start a chemical reaction with potassium permanganate crystals. This reaction involves rapid oxidization, generating great heat.

① Take a teaspoonful of potassium permanganate crystals and place them on a sheet of paper, cloth or some other inflammable material.

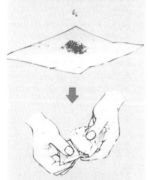

② Add 2 or 3 drops of antifreeze, and roll the sheet up tightly – this is vital in order to concentrate the heat in one spot, thereby raising the temperature to flashpoint (for paper this is 451 °F).

③ After a short delay – perhaps 1 minute – the mixture should ignite and set fire to the tinder.

An improvised tin stove.

Pyramid Fire Building a pyramid fire is simply a matter of placing logs in a pattern to create a pyramid-shaped stack. Smaller and drier twigs and sticks can be placed inside the fire or threaded between the layers.

This type of fire, once lit, will burn quickly and provide plenty of heat. It can also form the basis of a signal fire (see Signalling).

Improvised Stove
Constructing a stove from any available metal drum is a vast improvement on a simple open fire. A stove will save fuel, as it is 50% more economical than an open fire. With care, it can be

used inside a shelter, with the outer metal radiating enough heat to dry wet clothing while the stove provides light. If not too large it can be carried with you, complete with hot or burning embers.

Benghazi Stove If liquid fuel is available then a simple stove can be constructed by simply filling a large can half full of sand. The fuel is added until the sand is completely saturated. It is best to burn a small amount of tinder on top of the sand; this preheats the fuel and ignites the sandy surface, which provides a good slow-burning flame. Peat or fine gravel can be used if sand is not available. A fuel candle can be produced in the same way by using a smaller can and a strip of old cloth or rope.

Steps to Remember

1 Choose a suitable site and prepare it for the fire.

2 Gather an ample supply of fuel, grade it into categories and stack it.

3 Prepare your tinder.

4 Light the fire with small amounts of your driest fuel, and nurse it until it is burning hotly.

5 Add new material to it slowly – you do not want to smother it.

6 Check ventilation if there is any risk of carbon monoxide poisoning.

Start your pyramid fire quickly by building a small stack of feathered sticks and dry twigs in the centre.

Fire Management

• Don't make your fire so large that you can't get near it to put a pot on.

• It is safer and better to cook using only the embers.

• If you have the means to make one, an improvised stove is more economical of fuel than an open fire, and more versatile in use.

• Keep plenty of fuel handy, and near enough to the fire to dry out before use.

• Stop your fire spreading: it's dangerous, and wasteful. Beyond a certain size you get no additional benefit from a larger blaze – and your energy expended in gathering fuel is ultimately wasted in the sky above your fire.

• If you stay in the same place for any period of time, use the same spot for your fire.

• Keep drying clothes far enough away not to fall into the fire.

Next to air, water plays the most vital part in daily survival. The human body, which consists of roughly 90% water, cannot survive without water longer than three days in a hot climate and 12 days in a cold one.

Tropical rainfall is usually far higher than in temperate regions. In some areas rain falls on most days, often at a predictable time; it comes suddenly, and is followed by clear sunlit skies. In other zones rain is seasonal, and in the wet season falls continuously. This predictability makes it easier to plan your activities.

The tropical climate provides the perfect breeding ground for bacteria, viruses and parasites; it is very likely that any unprotected person entering these regions will contract some sort of disease. Although these are more prevalent in populated areas, where they are caused by lack of sanitation and clean drinking water, you should assume that all jungle water other than direct rainfall is contaminated, and should be purified – bad water can kill more quickly and painfully than no water at all.

Water is continually lost through the normal bodily functions of urination, excretion, breathing and sweating. The amount of water lost through sweating increases when in hot conditions or during physical activity. This water must be replaced.

Drinkable Water Sources

Even if you practice all possible precautions, without a good supply of potable water they will only prolong your survival by a few days. It is imperative to locate or extract water from any source available while being equally cautious about filtration and purification. Waterborne diseases and parasites pose a great health risk to the survivor.

Rain, streams and rivers provide the majority of the world's drinkable water, but it is not always easy to find.

Animals and insects will give some indication of water being present: watch grazing animals in the early dawn or at sundown, as this is when they will make their way to water. If surface water is not apparent, try looking in valley bottoms, and start digging. If neither surface nor sub-surface water can be located, you can acquire it from a host of sources.

Plant Sources Some plants have a high water content and offer a water source in emergencies but it is wise to make a study of plants in a particular area before you go. Bear in mind that some are poisonous; you should never drink from any plants with milky or coloured juices (two exceptions to this rule are coconuts, and the American barrel cactus).

Many jungle plants will act as water containers, catching any rain or moisture in the atmosphere. Try shaking hollow old bamboo stems to see if you can hear any water inside; if you can, pierce the stem carefully just above each joint and

A bromeliad.

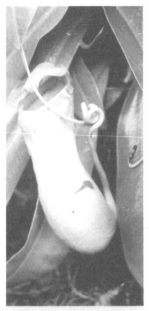

The pitcher plant, a good supply of water that should be boiled before drinking.

Slicing through vines provides fresh water.

let it pour out. Bromeliads, which are part of the pineapple family found in tropical America, collect water in between their leaves. Other examples of water-holding or collecting trees are the baobab of Africa and Northern Australia, and the umbrella tree of tropical West Africa.

Banana and Plantain Both these trees provide an ample supply of good drinking water. Cut the tree down, preferably sawing through the trunk 30cm (12ins) above the ground. Place your knife in the middle of the stump with the blade at an angle of 45 degrees, and hollow it out to form a bowl. You will find the water filling it even as you cut. When you have finished, scoop out any debris and wait for it to refill. Although sometimes bitter, the water is good. Cover the stump to protect it from insects, or use it as a means to catch them.

Vines Some vines may also supply fresh water. Cut deeply into the vine as high up as you can reach, then sever it completely near to the ground. Allow the water to drip from the end of the vine into your mouth or a container. Try a little first; water from vines is normally crystal clear and sweet. If the liquid is murky and causes irritation to the mouth, stop immediately as it could be poisonous. Once the water has stopped dripping, cut another section of the vine and start again. Study the vine so that you will recognize it in future.

Pitcher Plants Found all over South-East Asia, this is a climbing plant which is mainly confined to the more mountainous areas. Many people mistake the pitcher for a flower, but it is really a leaf formation which can be found in a variety of different shapes. Due to the poor soil the pitcher has evolved to supplement the nutrients it needs to sustain life by digesting the small insect and animal life that becomes trapped in the pitcher's water reservoir. Ground-level pitcher plants hold a good supply of water, but due to the plant's prey it is best boiled before drinking.

Ten minutes of jungle rainfall will deliver several litres of drinkable water if you stretch out a poncho.

Filtering Filtering will remove mud particles, leaves and small waterborne creatures. This can be done using a clean sock, a shirt-sleeve (or women's tights in the case of female survivors), a plastic bottle or a section of bamboo. Fill the makeshift filter with a layer of fresh grass or moss; then add either sun-dried sand or charcoal from an old fire. Allow contaminated water to filter through and run out the bottom end. Don't worry about the water being discoloured, especially if you are filtering with charcoal or are using peat water; this will do no harm.

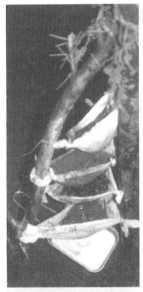

An improvised water filter.

Filter Hole A filter hole can be made in any form of waterlogged terrain such as a swamp, bog or marsh. Clear the vegetation and dig a hole above the water line, measuring approximately 30cm (12ins) in diameter and 30cm deep. The water which seeps into the hole may be dark in colour; this of itself is of no consequence, but the water will need boiling to kill off the micro-bacteria and viruses. If you do not have the means to dig, simply remove a large stone or log and let the well underneath it fill up.

LIFESAVER
Dangerous Water

Watching animals may lead you to water; but beware of drinking directly from stagnant pools where animals drink, as they transfer a variety of harmful parasites into the water through infected urine or faeces.

Leptospirosis, bilharzia, and dysentery are just a few diseases which can be caused by drinking contaminated water.

• **Animal bones in the vicinity of stagnant water may mean that the water is poisonous.**
• **Always sterilize or if possible distil water from ponds.**
• **Approach any isolated watering hole with caution, as it will be visited by a large number of animal species – some of which have large teeth and are permanently hungry.**

Sterilization and Distillation

Once you have filtered your water the next stage is to sterilize it. Sterilization can be achieved by boiling water vigorously for at least ten minutes. Make sure that the heat is distributed evenly – keep your water on a rolling boil.

Contained water, urine and seawater can all be made drinkable by distillation. This is a process whereby the contaminated water is converted to steam by boiling; the resulting steam is condensed and converted back into good drinking water. The process can be carried out with or without the aid of a fire, although some form of heat is required.

Another simple precaution is chemical sterilization, e.g. using chlorine-based purification tablets, potassium permanganate (see Survival Medical Pack), or iodine. Be sure to follow the instructions for use carefully.

Chemical sterilization tends to leave an unpleasant taste and odour in the water, and both the iodine and

Sterilizing Water

• Drinking bad water causes weakening sickness, and is more dangerous than thirst.

• Filter water first, using moss and charcoal.

• Kill off microbacteria and viruses by adding chlorine-based purification tablets.

• If you have no purification tablets, use potassium permanganate crystals.

• If you have no chemical agents, then boil water hard for 10 minutes minimum.

potassium will stain the water pink. Adding small pieces of charcoal to the water an hour before you want to drink it can rectify this.

Carrying Water

The survivalist should carry a supply of water even when travelling through an area where it is abundant. There is always the possibility that a lone survivor will fall or otherwise injure himself and be unable to walk. Any available container can be used, but those with a screw cap are best. Bottles, waterproof cloth, condoms, animal intestine and bamboo can all be fashioned into makeshift water carriers.

Salt

Avoiding Waterborne Diseases and Parasites

• Collect rainwater as it falls.

• Purify all drinking water, whatever the source, by filtration, chemical agents or boiling.

• Avoid bathing in untreated water.

• If forced to cross water, do so fully clothed.

Salt is next in importance to water, as it helps to regulate the fluid balance in the body. Without an adequate supply you will succumb to muscular cramps, heat exhaustion and heatstroke. The average human body requires about 10gm (0.35 oz) of salt daily to replace that lost in normal sweating.

When the body is deficient in salt, the first signs are sudden weakness, muscle cramps, dizziness, nausea and a hot, dry feeling all over the body. If these symptoms appear, rest and a pinch of salt in a mug of water are the quickest treatment.

Salt deficiency is common in tropical jungle conditions, so in these environments it makes sense to ensure that you add a small amount of salt to your drinks. It would also be a wise precaution to add some salt tablets to your personal survival kit.

Heavy-duty non-lubricated condoms will hold up to 1.5 litres of water supported in a sock or shirt sleeve.

Food is not an immediate
factor in survival. The
average adult can go 14
days without food before
any serious effects start to
impair physical ability, and
death from starvation takes
well over a month.

That said, all survivors should be on the look-out for food from day one; if it grows, walks, crawls, swims or flies it is probably edible.

Animals and plants form the two sources of food available from the wild. Animals provide food rich in energy, protein and many nutrients, but the survivor will usually have to expend much time, effort and energy to catch and prepare them. The amount and type of food you will be able to eat will also depend upon your water supply.

The jungle environment is rich in plant and animal life. Edible animals include monkeys, wild pigs, birds and rodents. Toads and salamanders can also be eaten, but have glands on their skin which need to be removed first. Fish are plentiful in the tropics, as are shellfish in swamps and coastal regions. Edible plants include wild bananas, mangoes, yams, oranges and lemons, breadfruit and taro.

Plant Food

Compared to the problems of catching animals, plant food is easy to gather and is usually available. The plant species will determine its richness in vitamins and minerals. Although some plants are very low in food value, they can still be sustaining.

In a long-term survival situation plant food on its own may not provide a fully balanced diet, and you may have to eat more than normal to fulfil your body's requirements. However, in times of need plants are a valuable resource and will keep you from starving.

Some knowledge of edible plants is required, as over half of all plant species are inedible or poisonous. Of those that are edible, only certain parts of the plant may be palatable. Whether you die of starvation or take the chance of eating a poisonous plant will be a personal decision at the time. If you choose the latter, you should at least take the precaution of doing an edibility test (see overleaf). Although not infallible this does give some indication of the human body's reaction to the plant.

Some edible plants contain elements that are dangerous to health if they build up in the body. Therefore be wary of eating too much of the same plant. A varied vegetable diet will not only be tastier but will also provide much more balanced nutrition.

The plants mentioned below are intended only as a guide. They represent only a small representative fraction of the plants which have uses as either food or medicine or both. It is recommended that you read about the plants that are native to the area where you intend to travel – learn to

Young shoots of bamboo are a useful food source.

LIFESAVER
THE EDIBILITY TEST

The edibility test is a time-consuming and thorough process. Although it may appear to be over-cautious, remember that your very survival is at stake. Plant poisons may take time before they have any effect on the body; also, plants may affect people in different ways. Make sure that you carry out plant testing before your food stocks are depleted, not after.

• A plant's identity must be 100% established. If for any reason you are at all unsure whether it is edible or not, follow the simple steps below.

• Be scientific and thorough in your testing. Test only one plant and one person at a time, so that any effects can be well monitored.

• The plant edibility test will **NOT work for fungi**.

• Avoid collecting plants from any area which may have been contaminated, and those with milky saps (except for dandelion, goat's beard and coconut).

• Wash any plant material thoroughly before cooking, and remove any diseased or damaged parts.

• Not all the parts of any one plant may be edible. Separate the root, stem, leaves and any fruit. Treat each part individually with the same test.

• Only test plants which are plentiful in your environment. There is no point in subjecting your body to possible poisoning if there is only a handful of the plant available.

The Test

❶ **First test the plant for any contact poisons. Crush a leaf and rub a little of the sap onto the sensitive skin of the inner wrist. If after 15 minutes no itching, blistering or burning has occurred, continue.**

❷ **Take a small portion of crushed plant and place it in your mouth between your gum and lower lip. Leave it for 5 minutes, testing for any unpleasant reactions.**

❸ **If there are none, chew the plant; note whether it exhibits any disagreeable properties such as burning, extreme bitterness, or a soapy taste.**

❹ **If it still gives no reason for suspicion, swallow down the juice but spit out the pulp. Allow 8 hours to pass to see if it has any adverse effects on the body, such as sickness, dizziness, sleepiness, stomach aches or cramps.**

⑤ If none of these symptoms occur, eat a slightly larger amount, e.g. a teaspoonful, and wait for another 8 hours.

⑥ If there are still no negative results, eat a handful of the plant and wait for a further 24 hours.

⑦ If after this period the plant has given you no ill effects, you can assume that it is safe and can be eaten in greater quantities.

Use all your senses

• As well as the taste test described, use your eyes; brightly coloured plants may be poisonous.

• Watch to see if other animals eat the plant.

• Smell may also provide you with clues to a plant's safety – be wary of plants emitting pungent odours.

recognize them and know their properties, in order to keep yourself and others safe.

Plants

Bamboo The young shoots can be eaten, although they have an extremely bitter taste and need to be well boiled, with a couple of changes of water. Bamboo flower seeds are also edible and should be cooked like rice.

Wild Yams Many species of wild yam occur in the tropics; they are perennial, and prefer light forest and clearings. Wild yams have twining, vine-like stems and a long, contorted rootstock or tuber, which is the edible part. Some vines also produce edible aerial tubers. It is best to peel and cook them before eating as a few varieties are poisonous in their raw state. Yams have a high nutritional value, being a good source of carbohydrate and vitamin C. They also store well as long as they are kept in a dry place.

Baobab (Adansonia) This bulbous-trunked tree is found in open tropical bush from Africa to Australia. It has a very strange appearance, the diameter of its trunk being equal to about half its height. It produces large white flowers which mature into pulpy fruits about 20cm (8ins) long. The leaves may be eaten, cooked in soups, and the fruits and seeds are edible raw. The swollen roots will, more often than not, contain a good water supply.

Wild Fig Most wild figs are edible, although some will be more palatable than others. Wild figs grow commonly in both tropical and sub-tropical areas. They all tend to be trees with a rambling habit and tough, evergreen leaves. The fruits, usually pear-shaped, may appear either in clusters or in pairs at the base of each leaf. Wild figs can be eaten raw and are highly nutritious, containing high levels of dextrose, calcium, potassium and iron, so should be considered as a very desirable survival food. Figs also have medicinal properties, being mild-ly laxative; they con-tain an enzyme which aids the digestive process. If they are roasted and split in two the inside of the fig can be applied with benefit to boils and abscesses, even those of the mouth.

Wild grasses grow in many parts of the world, even as far north as the Arctic circle.

Taro (Colocasia esculenta) The taro plant is a member of the arum family. It is found all over the trop-ics where there is moist ground, and is an important food plant for many native peoples. The plant can grow up to 1.5m (5ft) tall and has large,

The coconut is a good source for food as well as water.

leathery, heart-shaped leaves on stems rising from the base. It has an orange flower and a large turnip-like tuber; this tuber is the edible part but, as it contains harmful substances, it must never be eaten raw. The tuber should be cooked thoroughly; the result should be a root tasting rather like a potato. Taro roots are rich in starch.

Ti-Plant The ti-plant prefers to grow in the shade. It is a shrub which can grow to a height of 4.5m (15ft); the leathery leaves have thick stems and are usually green, although they may also be tinged with red. Although it bears red berries it is not these which form the edible part, but the fleshy roots. These are full of starch and highly nutritious; they are best cooked.

Clamping Many root vegetables can be preserved through the winter by 'clamping'. A thick layer (20cm/ 8ins) of dry straw or bracken is used as a base onto which the tubers are placed in a pyramid. Cover the pile with more straw or bracken; and allow it to settle for two days before covering the whole pile with dry earth. It is a good idea to allow some strands of straw to protrude through the earth so that your 'clamp' may breathe.

Sunflower seeds (above) and poppy seeds (below) can be eaten and also used to produce oil.

Seeds

All edible cereals are derived from wild grasses that produce heavy seed yields. A wide variety of grasses can be found in most regions, from the coldest tundra to all but the hottest desert. Although laborious to collect, the seeds will provide a basic food. They are best removed by simply gripping the seed head and pulling backwards so that the seeds fall into your hand. They can be collected in any improvised container, such as a hat or spare shirt. Once you have collected enough, rub the grain between your hands to loosen the chaff, and separate by throwing the whole lot into the wind, which will blow away the lighter chaff.

The seeds can then be ground into flour using a flat surface and round smooth stone. This flour can be mixed with either nuts or fruit, and baked into bread or biscuits.

Oil-Producing Plants

Sunflowers, poppies, olives and walnuts all produce oil which is both edible and can be used as fuel for lamps.

The poppy is one of the easiest flowers to recognize growing in the wild, and is almost always found in abundance. It favours recently broken ground. In moderation the seeds can be eaten raw with no ill effects, but they are best used to produce oil.

The seeds or fruit of any oil-producing plant need to be harvested and wrapped in cloth to make 'cheeses' (flat, round cakes); these are stacked on top of one another, and pressed. If using seeds, they are best cracked first on a smooth stone before being pressed. In a pure survival situation pressing presents a problem and some form of leverage needs to be implemented. A press can be made if a vehicle jack is available.

The residual 'cake' is also edible, and is best rolled into biscuits and fried. Olives can be wrapped in a clean cloth and left out in the sun. The oil exudes into the cloth, which can then be wrung out. The cloth can then be used for lamp wicks.

Nuts

Nuts are an extremely valuable food source and can be found in most countries and climates except for the Polar regions. Nuts are extremely nutritious, providing high levels of protein, fats and vitamins. Tropical nuts include coconuts, brazil nuts and cashews.

If you have a plentiful source of nuts, gather as many as possible and store them in a cool, dry place. The nuts will remain edible for several months if left in the shell. Nuts are quite easily carried and make an excellent portable food store.

The brazil nut is contained within a large fibrous husk.

Fruits

Fruits can be extremely high in vitamins and sugars, and often occur in plentiful amounts. Do not gorge yourself on wild fruits, however, as this may well cause severe diarrhoea and sickness. What you can't eat at once, collect and dry. Make sure that you dry them thoroughly, however; otherwise they will become coated with harmful moulds and mildew. For the same reason, only pick and eat fruit that are healthy and not overly ripe.

Fungi

Fungi provide a nutritious and palatable wild food source, and they often occur in areas where other food resources are scarce. Only two to three per cent of fungi species are poisonous to human beings; and yet opinion is divided on advising their use as a source of survival food. The major problem arises from the fact that **THERE IS NO EDIBILITY TEST FOR FUNGI.**

This is due to the delay between poisoning and symptoms appearing, and also the exceptionally toxic properties of some species. Even species not considered poisonous can cause some extreme reactions in susceptible individuals who may have an allergy to them. Just because one person can eat a certain species quite safely does not mean that every member of a survival party can.

- Even though only a small percentage of fungi are poisonous

The Danger of Fungi

- Only a small percentage of fungi are poisonous to humans, but these are extremely deadly.

- Death can occur even if only a tiny portion is consumed.

- NEVER try testing fungi for edibility.

- Even less poisonous species, not usually fatal, may cause life-threatening sickness and weakness in a survival situation.

to humans, some are extremely deadly – even if only a tiny portion is consumed.

- After eating a poisonous species of fungus the symptoms may not present themselves until ten to 40 hours later. By this time they will be serious enough to warrant hospitalization. In the worst cases, without hospitalization the casualty will die. Even with proper medical care irreversible damage may be caused to certain organs.

- Less poisonous species, although not fatal under normal circumstances, may cause poisoning serious enough to threaten the life of an already weakened person in a survival situation.

Identification of fungi

There is only one safe way to identify fungi, and that is by visual means. Make sure that you can positively identify certain species beyond doubt. Learn fungi identification by studying pictures from a good guidebook and comparing them to specimens found in the wild. Keep in mind that even reference books may disagree from time to time on whether a particular species is edible or poisonous. A decision to eat fungi must be based on sound know-ledge and first-hand experience of identifying edible species, and where no alternative food source is available in the locality. If you are not able to do this, it is safer to leave fungi well alone.

Preparation

All plants and leaves should be washed in fresh water before consumption. While some of them can be eaten raw, it is generally safer to cook all food. Add plants and berries to other dishes, such as stews and soups. Not only will the addition enhance the taste of the food, but it will also add to the general nutritional values of your cooking.

Roots and tubers can be boiled, but they are much better baked or roasted. Wash or scrape first.

Leaves, stems and buds are best boiled until they are tender. Keep replacing the water if your plant material has a bitter taste – this will reduce it.

Grains and seeds can usually be eaten raw, but most will taste better after being parched. Parching is usually done in a metal container, but can be done on a hot, flat stone on top of a Yukon Stove if no other container is available. Heat the grains or seeds slowly until they are well scorched.

Nuts are in most cases perfectly edible raw, but some, such as acorns, can be very bitter. These are best boiled for two hours and then soaked in fresh water for three or four days. All nuts can be ground up into a paste, which can either be added to soups or stews, made into a gruel, or dried into a 'flour' to make unleavened 'bread'.

Fruit berries and soft fruits provide a welcome change to a survivor's diet. Both are best eaten raw as they contain many valuable vitamins which may be lost when cooked. Those fruits with thicker or tougher skins can be boiled, baked or roasted.

Animal Food

Mammals, birds, fish, reptiles, crustaceans and insects are all sources of animal food that can be found in the wild. Animal foods of any type will provide a higher food value than that obtained from plants; however, far more energy-sapping effort will be needed to catch an animal than to gather plants. Hunting and trapping all require time, skill, and good information. It is vital, therefore, that the result matches the cost of the methods employed. You must not expend more energy in catching the food than that derived from the food value caught.

Being able to observe animal signs is a valuable skill in the jungle. Despite the dense foliage signs are easily found in mud, streambeds, on trails and near watering holes. When animals move it is usually for a reason, and on a daily basis that reason is either to eat, sleep or drink. Tracks will give you information about their direction of travel, and an indication of their size.

Hunting Hints

Hunting can be achieved by either trapping or pursuit. The first requires constructing some form of trap best suited to catch your animal; pursuit means to stalk or ambush an animal and kill it by direct means, i.e. stabbing, clubbing or shooting.

Traps can be constructed to catch just about any size of animal from a mouse to an elephant. If pursuing your prey or waiting in ambush you will need a weapon; this can range from a rock to a gun. A firearm will provide the best chance of successful hunting, with snares coming a close second. The construction and efficient use of primitive types of hunting weapon require a great deal of skill and practice.

Unless you are an expert, hunting with anything less precise than a rifle will probably produce little success; but lying in ambush will increase your chances.

- To be able to ambush your prey you will need to know where it lives and when it moves. Look for an animal trail, especially one that leads to water. Most animals will use these trails between their feeding and bedding grounds and their water source either in the early morning or in the late evening; so pick one as your time of ambush.

- Animals have more acute senses than humans, and are always on the alert for danger. Be patient; observe all potential prey; camouflage both your appearance and scent – daub mud over your face and hands. Keep a low, silent profile and use smooth, careful movement down-wind while the animals are feeding. Find a good place to hide, and position yourself there well before any prospect of animal movement.

- Snares and nets work well when set around an area where an animal has been cleaned or butchered. The entrails will act as a very effective bait.

- Care must be taken when returning to a trap or a snare, as any wounded animal may be dangerous.

- A sharp whistle can stop rabbits and hares if startled into running. You may even be able to attract them to you by making a high-pitched kissing sound with your lips on the back of your hand to simulate a squeal.

- Birds should be watched to see if their movement will reveal a nest site containing nutritious eggs or young. These should never be overlooked as a food source.

- As a last resort, the survivor must consider eating anything that walks, flies, swims, crawls, creeps, jumps or wriggles.

Author's Note

It is my firm belief that there is no justification for hunting any animal for sport. Only within the context of this book do I advocate hunting, and then only as a matter of human survival under the laws governing nature – in as much that the strong and intelligent of any species will survive by preying on the less fit. Even in this situation the hunter must act responsibly, and not let any animal suffer unnecessarily. (All the dead animals pictured in this book were purchased dead from local country markets.)

- Where possible, try to use all of a carcass – do not discard anything without careful thought. Skins can be made into clothing; bones can be fashioned into arrowheads, fish hooks or needles; sinews and gut make good bow strings or sewing thongs.

Snares

Snares and traps are a far better alternative to hunting and ambushes, as they require less physical effort and time spent waiting. A well-made and correctly sited and set snare or trap will be effective 24 hours a day, without the need for constant vigilance. This method guarantees a 'cost-effective' meal in terms of the effort/benefit equation of survival. Start out by snaring small game; they are easier to trap, transport and prepare.

Always set several snares, but keep some distance between them; an animal caught in one snare will create enough noise to alert others to the possibility of danger. Make sure that all snares are checked on a daily basis – the caught animal may be your next meal, but there is no reason to let it suffer unnecessarily. If you are successful with any of your snares, collect the animal, kill it if necessary, and reset the snares for the next day.

Drag Snare If properly positioned the simple drag snare is a most effective way of catching a meal. Ideally the snare should be placed along a fresh run, in such a way that the animal's head will be caught. Tie the noose to a stake which has been driven firmly into the ground; or, if it is suspended above the run, secure it to a strong branch. To set the noose, position it so that its bottom edge is about 10cm (4ins – the width of the average hand) above the

A drag snare.

floor of the run; and adjust the noose until it is about the size of two clenched fists. If possible, encourage the rabbit or other animal into the trap by piling up dead twigs and branches on either side of the path leading up to the snare. However, do not make the mistake of using green twigs – the animal may consider these to be a distractingly tasty snack.

Balanced Pole Snare The simple drag snare can be modified to make it even more efficient. A balanced pole snare will not only catch your prey, but will also lift it clear of the ground – out of reach of any other hungry predators or scavengers. This requires a suitable length of pole secured at its mid-point across the trunk of a nearby tree in such a way that the lighter end of the pole can be pivoted downwards directly above the animal run. Fix a snare firmly to that end of the pole. At the opposite end of the pole fasten a heavy rock to act as a counterweight. This counterweight should be heavy enough to lift your catch clear once the snare is activated.

Make a trigger by cutting inter-locking notches in two pegs, which hold them together against straight line tension (i.e. when you try to pull them apart along the axis of the pegs) but which slip apart easily when disturbed sideways. Hammer one peg firmly into the ground at the side of the animal run. Attach your snare to the other, free peg; and also tie a line from this free peg to the light end of the pivot pole. Swing the light end of the pole down and hook the trigger halves together; check that they work smoothly. Check

A simple trigger construction.

Figure Four Trap

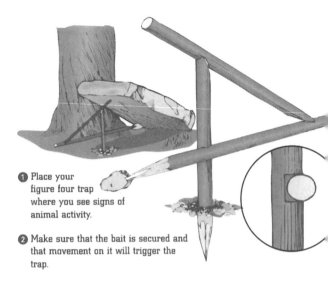

❶ Place your figure four trap where you see signs of animal activity.

❷ Make sure that the bait is secured and that movement on it will trigger the trap.

that you have set the noose at the correct height above the run and that the wire loop can move freely.

Spring Branch Snare A similar effect can be achieved by substituting a springy, bent-over branch from a nearby tree, or a bent-over sapling, for your pole and counterweight. Again, hold it bent down under tension by tying it to your notched trigger pegs and tying the snare to the free peg. If you intend using this method you are advised to check the spring strength of the branch beforehand, and adjust your trigger setting accordingly.

Hoop Spring Snare A hoop spring snare can be used where trees are scarce and you are forced to rely on small saplings. Using two saplings – either growing naturally close together, or cut down and firmly planted in the ground where you need them – bend them into an arch. The two tips are locked together by a notch which in turn is held in place by a vertical bait bar. (A rock can be attached to the bait bar if necessary, to supply the downwards tension to lock the notch.) A number of snares are attached to the saplings and positioned in such a way that the animal must pass its head through a loop in order to get at the bait. Movement on the bait bar will trigger the trap and snatch tight the snares.

Whore Trap The whore trap relies on forcing the animal's head into a baited 'V'. A willow stick, sharpened at each end, is bent into a hoop and forced into the ground. The snare is fixed to the end of a bent-over sapling, or the end of a balanced pole snare. A bait stick is positioned so that the snare peg, which fits through the hoop, can rest on it. Two large logs or a series of stones form a barrier either side, forcing the animal to place its head through the snare before it can eat the bait. As the bait is taken the snare is activated. Of all the snare traps shown this is by far the most reliable.

Purse Net A simple purse net, if you have one, is another efficient way of catching small game. If you do not have one, make a gill net. The net can be used in several effective ways. First find a burrow showing signs of recent use, and stake the net over a fresh entrance. Block all of the other burrow holes except for one. In this hole either light a fire and blow smoke, or simply pour in water. Either method will make any occupants of the burrow panic, forcing them into the net.

Long Netting This is a simple and effective way of catching several rabbits at once. You will require a long net, which is placed between the burrows and the grazing ground. It is erected rolled up and balanced on several sticks; a cord is attached which allows the net to be drawn out. It is best used after dark when the rabbits are feeding. Stretch out your net, and then get behind the rabbits and make a lot of noise. The rabbits' first reaction is to bolt for their burrow.

Figure Four Trigger This type of trigger has the advantage of being easy to make, light to carry around with you, and capable of supporting any combination of useful traps. It is constructed from three lengths of thick branch, notched in such a way that they form a figure four. This trigger is firmly fixed in the ground where it will support a deadfall log or flat rock, or alternatively a drop net. Whichever method you choose, once the trigger is disturbed the trap will activate. The Figure Four Trigger is the ideal trap to use while travelling.

Birds

All birds and birds' eggs are edible. Their taste depends on their habitat: those which live or feed at sea will be less palatable than those that feed on the land. The flesh from sea birds is nutritious but barely digestible, though this can be improved by thoroughly cooking it.

Snaring

The use of snares is discussed here purely in the context of survival; snaring animals is against the law in some countries, and is disapproved of in many others.

Making a Snare

The easiest type of snare, both to make and to use, is the drag snare, which kills by strangulation. A noose can be fashioned of any strong wire, nylon cord, hide strips, or even a wire saw (see Survival Kit). The best material to use is brass snare wire. You will need about 80cm (30ins) of wire for each snare. Make a 1cm (1/2in) loop in one end, passing the other end through the loop to make your noose. The pliability of the brass wire makes for a quick, smooth strangulation, which will lock in place as the animal struggles. Before setting it make sure that the wire is free of kinks and that the noose runs freely. Snares are best rubbed with animal excreta to remove the brightness of the metal and the human scent which your hands will leave on the wire.

The normal indication of bird presence is simply to see them flying overhead, but many also leave signs near their nesting or feeding areas. Although it is difficult to identify a particular bird species by its track, you can still get a rough idea of the type of bird. By using the following simple guide lines you should be able to tell the difference between perching birds, swimming birds and wading birds:

- Perching birds (e.g. sparrows & crows) leave tracks with a long first toe (the gripping toe) behind three front toes.

- Swimming birds (e.g. ducks) leave webbed footprints.

- Wading birds have long slender toes spread wide apart. You will find their tracks in mud.

Bird Snares Birds can be caught in any number of ways, from throwing a stone to hitting them with a long stick. One of the simplest ways is to snare them. First find a perch that is well used by birds – this can easily be identified by the large amount of droppings either on the branch or on the ground below. The snares can then be hung above this branch. Once a bird has put its head through a loop it will not withdraw but will try to escape by flying forward, and thus become trapped.

Another method is by using the baited perch. If you have sufficient wire – at least 2m (6.5ft), make a snare loop at either end and fold them over a branch. Next form a square-ended perch with the trailing end onto which the birds are enticed to land. When a bird rests on the perch it will dislodge the whole snare, trapping the bird's neck at the same time. In most cases both bird and trap will fall to the ground.

Baited Bird Hook A simple baited hook (an open safety pin is ideal) can be used to catch larger birds such as seagulls, wild ducks and geese. These birds are greedy and swallow their food quickly. Make sure the line is well secured, and that you check all of your snares each day.

Bottle Trap Floating traps can be used to capture waterfowl while on the water. If you do not have a bottle use a small log instead. Half fill the bottle with water, and tie two or three snares to the neck so that they sit about 5cm (2ins) above the water. A little foliage will make the trap more attractive to any curious bird.

Unless the water is shallow and safe enough for you to wade in and retrieve the trap, secure it to the bank with a line so that you can pull in any catch.

Eggs Any survivor should keep an eye out for birds' nests; eggs offer high nutritional value, are convenient and safe, even if the embryo has developed inside. They can be boiled, baked or fried. Hard-boiled eggs can be carried as a food reserve, and if submerged in clean water will keep for several weeks. A thin coat of fat or grease around a fresh egg will keep it edible for a month or more. A survival diet of birds' eggs and boiled nettles will sustain life for a long time.

Bird and Fish Catcher

In isolated regions where man is rarely seen, most birds will remain perched and unafraid. Use a long gaff with a snare attached to hook your dinner.

Never remove all the eggs from a nest; by leaving one or two you will encourage the bird to lay more. Mark those you leave, to ensure that you are removing only the fresh eggs.

Preparation of Birds

Before cooking, birds need to be prepared by plucking and cleaning. Most birds can be plucked more easily either immediately after death, or after being plunged into boiling water. The exceptions to the latter are waterfowl, which are easier to pluck dry. Do not throw away clean feathers as these can serve many purposes, from insulation in bedding or clothing to making flights for arrows. Although it is possible to skin a bird, removing its feathers at the same time, remember that the skin will provide extra food value.

Once the bird has been plucked, cut off the head and feet and make an incision into the lower stomach below the breastbone. Use this hole to draw out the bird's innards and neck bone. (The heart, kidneys, liver and neck bone will form the basis of a good stew.) Wash the bird thoroughly, both inside and out, with fresh water. Small birds, once gutted and cleaned, can be enclosed in clay and baked on an open fire; the feathers and skin will pull away with the clay.

Birds are easily trapped by rigging collapsible perches.

Carrion eaters – e.g. vultures, buzzards and carrion crows – are likely to be carriers of disease and parasites. They are still edible, but need to be boiled first for at least 20 minutes before you continue with any other form of cooking. Boiling will not only kill any parasites and bacteria present, but will also serve to make stringy meat more tender.

Traps for Larger Game

Scissors Trap A simple scissors trap features one log raised above another in a V-shape. The falling log is held in position by a trigger, and the direction of its fall is guided by stakes. It is essential that both trigger and retaining cord are strong enough to support the deadfall, yet upon activation will release quickly and smoothly. The falling log can be weighted to improve kill efficiency. One of the best trigger release systems is where two pegs or modified branches support a toggle attached to the release line. The toggle itself should be baited to avoid the risk of the cord being chewed by the animal, and placed in a position where the animal must expose its neck in order to get at the bait.

Deadfall and spears.

Deadfall and Spears A variation on the scissors trap is to cross the animal trail with a trip line, which when activated will drop either a log or weighted spears. *Note:* Many survival books illustrate this trap with the deadfall or weighted spears falling or swinging across the line of the path. Situating the fall to activate along the line of the animal trail will produce much better results.

Baited Pit Constructing a trap by digging a hole takes a lot of energy, although there are times when the ground is soft and the surrounding area is habitat to the ideal catch. The jungle is just such an environment, and wild boar and pig the game.

Providing you have the means, you need to dig the pit at least 1m square by 1.5m deep (3.25ft square and 5ft deep). Placing sharpened bamboo stakes in the bottom may help disable the animal, but they are unlikely to kill it. Covering the pit so that it matches in with the natural surroundings is vital. Likewise, the support for the concealing cover needs to be firm enough so that it gives way only when the animal is 'centre stage' – this can be achieved by cutting part way through the supporting branches. Always approach an activated pit with care: injured animals can leave a nasty infected bite. Make sure your prey is dead by stabbing it with a spear before attempting to remove it from the pit.

Bait The use of bait will increase your chances of catching a meal, be it an animal or fish, but what you use as bait is important. The idea of baiting is to attract the animal by offering an easy meal, and to optimize the efficiency of the trap or snare. In the first instance the bait must be acceptable to the animal; there is little point in using a worm if the wet ground is covered with them. Conversely, strange-looking bait may make the animal wary. Almost all animals and birds are attracted to blood, brightly coloured berries, and salt.

Hunting with a Weapon

Most improvised weapons, whether hand-held or projectile, require the addition of a cutting or piercing blade or edge. These can be fashioned from a wide variety of materials. Stone can be chipped to form an edge, and flint is particularly good for making weapons. Shave wood with a knife into a point and harden by charring slightly over a fire. Some woods, like bamboo, are naturally hard and only need trimming to a point. You can use man-made materials such as metal and glass to produce a good cutting edge.

AR-7 Survival Rifle Though it is rare nowadays to find a rifle packed in a survival kit, they do exist, and in certain environments prove most useful for hunting. Most survival weapons are of small calibre, since the relatively devastating ammunition used on the battlefield is unnecessary for hunting. The popular AR-7 survival rifle fires a .22 Long bullet. The AR-7 conforms to the needs of a survival situation, since it packs down for carriage into its own hollow stock, is lightweight, and will even float in water. Its 20-round magazine should, if used with sensible economy, supply sufficient meat to last several months.

The weapon is semi-automatic, which means it will fire a round each time you pull the trigger, i.e. you are not required to cock the

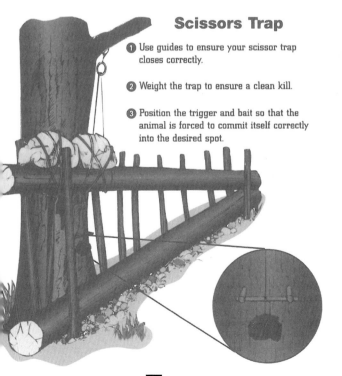

Scissors Trap

1. Use guides to ensure your scissor trap closes correctly.

2. Weight the trap to ensure a clean kill.

3. Position the trigger and bait so that the animal is forced to commit itself correctly into the desired spot.

weapon each time. For this reason, be careful not to let your trigger finger 'run' – aim for one round, one kill. Try to shoot an animal that will provide a good amount of meat, such as a fox, wild pig or capybara. Rabbits and birds can be caught by snare and are a waste of ammunition. Conversely, if you hunt game that is too large, such as a moose or bear, you will only wound it – which is wasteful of ammunition, cruel, and often extremely dangerous.

Assembling the AR-7 is simple:
- Open the rear of the stock, and empty out the parts.
- Slot in the trigger housing and bolt action assembly.
- Match up the barrel and body notches and secure with the screw collar.
- Check all parts are hand-tight; then fit the magazine.

Zeroing Under survival conditions ammunition may be limited to one full magazine (20 rounds). If the rifle is inaccurate, you could miss with every shot. You are advised to test the rifle by firing three rounds at a large target.

From a distance of 50m (55 yards), fire at the same fixed point each time. Estimate an imaginary point at the centre of your three bullet holes, and measure the distance and angle from your fixed point. If the centre of your group is left 5cm (2ins) and slightly high, you need to aim off to the right by the same distance and slightly low. Aiming off is better than adjusting your sights, as you will need to confirm any adjustment by firing more ammunition. Always aim at the centre shoulder area of the animal.

Balala Light

'Balala light' is an African term for hunting with a light. A powerful torch is attached to a helmet or hat and aligned with the eyesight. When game are near the light is switched on, illuminating both the animal and the gun sights. The animal is temporarily frozen by the bright light, and is easily killed.

Bow and Arrow Providing you can locate the correct materials it is possible to make a good hunting bow in a fairly short time. The most important part of the process is to select your stave – the part that forms the arc of the bow. Select carefully, choosing a strong, healthy section of wood without side shoots. The most traditional wood to use is yew, but oak, birch and hickory are all suitable. The wood should be long enough to make a bow stave about 130cm (50ins) in length.

Flex your stave several times to find which side bends naturally. Mark this side, and taper off the last 50cm (18-20ins) at both ends. Traditional English bow makers always tapered their bows to a round section and made the ends as even as possible – this was to stop the bow twisting when it was drawn. The bow stave should be slowly dried over a fire for about two or three days. Notch the ends to receive the bowstring.

To string the bow use whatever strong cord you have to hand; parachute cord will do. One alternative is to use cleaned animal

Making a Bow

1. Select a hardwood staff about 1.3m in length that is free of knots and limbs. Chamfer a third of the length at each end.

2. Notch both ends to receive the bow string.

3. String the bow by securing one end and forming a slip-over loop on the other.

intestines dried and twisted together to form a string. Tie the string on to one end of the bow only. Make a loop in the other end so that it can be slipped over the other end when the bow is flexed. The bow should only be strung like this when you intend to use it; at all other times it should be left untensioned.

Stringing a bow.

Arrow shafts Arrows are made from straight, strong wood about 65cm (25ins) long and 1cm (0.4in) in diameter.

Most types of wood will do, but choose birch saplings if you can find them. Clean any bark off the arrow and straighten it as much as you can – a good method is to gently chew the arrow between the teeth. Remember – a straighter arrow will fly further and hit with greater force. Balance the arrow on your finger at its halfway point. Insert your flight in the lighter end and the arrowhead at the heavier. At the flight end cut a notch about 6mm (0.25in) deep to take the bowstring – check the width against your string material.

Arrow flights Arrows need flights – 'feathers' – in order to keep them on course when shot. They can be fitted with double or treble flights, and these can in practice be made from actual feathers or plastic, polythene or cardboard. The flights should be 10cm long and 5cm wide (4ins by 2ins). In a survival situation a

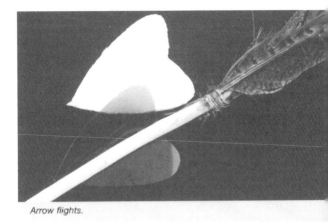

Arrow flights.

one-piece double flight is best used.

Using a knife or other thin blade, carefully make a split about 15cm (6ins) long into the flight end of the arrow shaft. Insert a double flight into this – i.e. a single piece which protrudes equally on either side of the shaft. If the arrow splits completely, bind the split ends together tightly with light cotton, fish line or very fine snare wire.

Arrowheads A variety of arrowheads can be made using different materials, but all are attached in a similar manner to the flights: carefully split the shaft, insert the head, and bind the split shaft tightly. If you can find nothing to act as an arrowhead, harden the tip of the shaft by turning it slowly in a fire. Once hardened, any charred material should be removed and the tip sharpened to a point.

Blow Pipe Although most people think of the blowpipe as a weapon used by jungle tribes, it is possible to construct a very effective modern-day variation which can be used for hunting small game such as birds and rabbits. Most of the materials required can be found in any modern vehicle or aircraft. For example, the body of the blowpipe can be constructed by simply cutting out a length of fuel pipe. Choose a section that is straight and at least 1.5m (5ft) in length; if this is not possible, try joining two or three shorter sections together. More air will be forced down the pipe if a mouthpiece is fitted at one end; this can be cut from card or plastic and held in place with ducting tape.

Metal darts between 10 and 15cm (4-6ins) long are constructed from stiff wire. Heat one end in a fire until it is glowing red, then flatten it to form a point by beating. Allow it to cool or dip it in water. The flight can be made from any soft, pliable material, e.g. seat foam or polystyrene. Use a short

Improvised arrowheads.

section of pipe which has the same diameter as your blowpipe to stamp out your flights; this will ensure an airtight fit, while allowing the dart to be blown easily through the blowpipe. Shooting with your blowpipe needs no explanation, other than to say that your accuracy will become second nature after a little practice. The example illustrated here has a range of 25m (80ft), and is capable of killing a rabbit.

Slingshot The slingshot is a very simple weapon, easy both to make and, with practice, to use. Take two equal lengths of cord or leather about 35cm (14ins) long, and attach one end of each to a small, shallow pouch of fabric or leather which will hold a walnut-sized pebble. Tie a loop in the opposite end of one cord, and a knot at the end of the other.

Place the loop over the index finger of your dominant hand, and trap the knot between index finger and thumb. Place your ammunition securely in the centre of your pouch – ideally this could be a small, smooth pebble. Bring the sling above your head in one quick swinging motion to gain momentum. Let go of the knot to release the stone. You do not need to swing the sling more than a couple of times. Try using a flicking action to improve accuracy.

Throwing Stick Used properly a throwing stick is a most effective means of knocking down and stunning a running animal. It is best to cut several 50cm (20in) lengths of heavy fist-sized sticks for throwing.

A blowpipe in use.

The Bolas

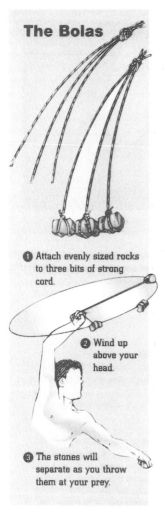

❶ Attach evenly sized rocks to three bits of strong cord.

❷ Wind up above your head.

❸ The stones will separate as you throw them at your prey.

Hurl them overhand or by side-throw, using a flicking motion on release to make the stick spin through the air. Advance on the animal the moment it is down, and club it to prevent undue suffering.

Club The club is probably rivalled only by the picked-up stone as the oldest known weapon. In its basic form it will extend the range of you arms and deliver a more powerful hit than your fist. Clubs can be made from either wood, stone or metal, and can be weighted or formed into a 'mace'. Construction of any club should be designed around its planned use and the ability of the user. Making a club will protect the survivor against some larger animals, such as wild dogs; and will also serve to ensure a clean kill of any animal caught but struggling in a trap.

Bolas The bolas is a very effective weapon for bringing down large, long-legged animals such as deer, wild sheep or ostrich. It is simply made, comprising three lengths of strong cord knotted together at one end and weighted with stones at the other ends. The stones should be of even weight and no larger than a duck egg.

Practise on a nearby tree by swinging all three lines above your head. When you let go of the knotted end the lines will separate and wrap around your target. Be ready to spear your game the moment it is down, as the bolas will not immobilize it for long.

Catapult If you have the means to make a catapult under survival conditions it will prove to be a highly effective hunting weapon. All you need is a strong, forked twig and a length of elastic (you might even consider putting some into your survival kit). A good source is the rubber taken from a vehicle's inner tube. Avoid clothing elastic, as this is generally too weak for the purpose. Construction is simply a matter of tying the ends of the elastic to the forks of your Y-shaped twig and the other ends to a good-sized projectile pouch – tie them tightly, and make sure the pouch is centred.

If you have a good length of elastic available, try using an arrow instead of a stone. Once this method has been perfected you will find it both more accurate and more deadly.

Spears Spears are useful for protecting yourself against an attack by a wild animal, but they are of less use for hunting. To

Target Practice

To become accurate takes practice, and arrows take a long time to make. It therefore makes sense to practice shooting at a target that neither allows your arrows to get lost if you miss, nor breaks them if you hit.

make an efficient throwing spear and achieve consistent accuracy demands skills of a high order. A thrown spear is less accurate and projects less killing power than an arrow. For hunting a spear can really only be used against cornered prey, although fishing spears are of more value.

To make a spear, choose a strong staff about 180cm (70ins) long and sharpen the end. If you have the materials and the time, experiment by making spearheads from other materials, such as flint, or metal or glass from a vehicle or aircraft.

Spears

❶ Spears with multiple barbed heads are best for fishing.

❷ Split the shaft to attach a metal or stone head.

Animals to Hunt: Tracks

Whether or not you actually see a prey animal, you can try to identify and locate it by studying and following its tracks. Efficient tracking is a highly sophisticated skill, and acquiring it in a survival situation will present a considerable challenge to most people.

One animal's track can look like that of a completely different animal depending on the surface on which it is imprinted. Sand, mud and snow will alter the image of an animal's footprint. You will rarely find a perfect print with the elements against you. Snow thaws and rain will wash away mud, resulting in a distorted shape. Even if you are convinced that you are trailing one type of animal, it could still turn out to be another. There is always a chance that a young animal could leave a print like a smaller creature. The different tracks made

LIFESAVER

AVOID using your knife as a spearhead unless there is a very good reason to do so. You are liable to damage your blade – or, worse still, lose it altogether. Its value to you in any survival situation is far greater than that of a spearhead, which can be made quite easily from chance-found or naturally occurring materials.

Tracks

- Study your environment at length, and use your common sense. What animals are likely to leave tracks here? Where are they going, and why?

- How is the surface – sand, mud, etc. – affecting the tracks?

- Is the weather affecting the tracks?

- Is this a full-grown small animal, or a young larger animal?

- Can you tell the front from the rear prints?

- If so, do the tracks tell you anything about the speed of movement?

- Are there any other tell-tale signs – droppings, or chewed vegetation?

by fore feet and hind feet can trick you into thinking you are following the tracks of a different animal.

With all this against you, you need to have a clear idea of what to look for in the first place. Your conclusion should not be based purely on the print, but also on your surroundings. You should be considering what type of surface the track has been made in; the time of day or night; the weather conditions which may have affected the print; and, most important of all, the probable game in your particular surroundings. There are generally other clues, too.

As an example of similar tracks, consider a rabbit and a squirrel. The rabbit will push off with its hind feet and land on its fore feet, which touch the ground one after the other. The hind feet then touch the ground landing in front of the fore feet. This leaves a print of the larger hind feet, followed by the print of the smaller front feet. The squirrel has a similar type of movement, with the hind feet landing in front of the fore feet, leaving the same type of print. It would be very difficult to decipher which footprints had been left by which animal – if it were not for one simple clue. A squirrel's trail starts and ends at a tree.

A series of tracks will give you a trail which gives you an idea of the speed at which an animal was moving. The greater the gap between the groups of tracks the faster the movement. A walking animal moves its right fore foot first, followed by the left hind foot. Then the left fore foot is moved, followed by the right hind foot, and so on. A trail made by a walking badger will show that the hind foot has landed on the track of the fore foot. This is called 'being in register', and is what happens when an animal has been walking or trotting – it moves its legs in a definite order. If an animal has been galloping, the tracks will not be in register.

Rabbit Rabbits deserve a special mention; they are a great source of wild food, and are found on every continent living in all conditions, from Arctic to desert. They are easily recognizable, and being a social animal are always found in large numbers. They usually stay in one territory all their lives, where they live in burrows, often with more than one entrance. They are most prevalent in open grassy areas and open woodlands, especially where the soil is dry and sandy. Burrow sites are made in banks and slopes with light tree or shrub cover.

The tracks that rabbits regularly use are called runs. These are easily seen between the burrow entrances and, if in present use, will have rabbit droppings on them – small, dark, round 'currants'.

When you set a snare make sure that it is a little distance from the burrow entrance itself – animals are far more wary when emerging from underground than at most other times, and a snare set too close to the entrance may well be seen by the rabbit and avoided. Take care not to disturb the ground or foliage around the run when setting the snare, and conceal your scent by rubbing the snare and your hands with animal droppings.

A live rabbit is best killed by holding its hind legs in your left hand and its neck in your right. Stretch and twist the neck sharply until the neckbone breaks; death will be instantaneous.

Deer Deer can be found from the lower Arctic to the lower reaches of the jungle. They walk on two toes, leaving a definitive track. They live in open country and woods. Many have branched antlers, which they drop after the October rut – the mating time for deer and other hoofed animals. Red deer start off as spotted calves; during the summer their coats change to a red/brown colour. Their diet consists of grass, fruit, heather and tree bark; it is also not unknown for them to raid crops.

Rodents Rodents belonging to the subspecies known as myomorpha make up about a quarter of all mammals. The best known of these animals are the various types of rats and mice. They have adapted themselves to surviving in almost any location except for Antarctica and colder regions of the Arctic. Their diet consists of seeds and other vegetation; but certain species have become omnivorous, and will eat any food left out by humans.

The problem for the survivor is that rodents are the carriers of many diseases – leptospirosis, rabies, ratbite fever, murine typhus, bubonic plague, hantavirus and spirochetal jaundice, etc. Through their urine, droppings and hair food can easily become contaminated, and at the very least will pose a threat of bacterial food poisoning. Despite this, the animals are edible. This makes them a ready source of food, and one which mankind has often turned to, especially in times of famine.

Trapping vs. Hunting

• Hunting demands practised skills – silent movement, concealment, reading the natural environment, predicting animal behaviour. Survivors from urban backgrounds rarely have them. All potential prey animals do.

• Trapping demands the ability to visualize basic mechanical principles, to fashion simple materials, and to study the surroundings. These are skills which even urban adults can master well enough to deceive most animals.

• Hunting means movement, sometimes over long distances. This expends the survivor's energy. If he is unsuccessful, it is not replaced.

• Making and setting traps and snares demands little strength, and less movement across country – therefore less energy loss.

• The hunter normally has to focus on a single prey. If that prey escapes him, his time and energy have been wasted.

• The trapper can set many snares, all of which are potentially working for him simultaneously and for 24 hours every day. They are dramatically more productive by the equation of cost against possible rewards.

Monkeys Huge numbers of a wide variety of monkeys live in all types of rain forest. Jungle tribes regard monkeys in very much the same way as Western people see rabbits: as a food resource. Larger species should not be hunted, as the males can be extremely aggressive. Monkeys can be caught in traps or shot with a bow or blowpipe. Traps vary; the baited perch type is one of the best ways of catching a monkey, with bananas being the favourite food. The bait ensures that the monkey is in the correct spot when the spear-laden arm is activated. The trap should be designed to keep any catch from being pilfered by other predators, as a monkey's scream can be heard for a great distance.

Mousedeer The world's smallest hoofed animal, the mousedeer is no larger than a domestic cat. Although found throughout the Far East its main habitat is the Malaysian jungle. It is an extremely shy animal and rarely seen during daylight. The best way to catch them is to search with a torch while they feed after dark; caught in the torch beam, a mousedeer will normally freeze long enough to be felled with a club.

Giant Capybara The largest of the rodent family, measuring about 60cm (2ft) at the shoulder and weighing up to 45kg (100lbs), this herbivore lives in large groups along the banks of South American rivers. It is semi-aquatic, usually diving into water at the first alarm. The coat is generally short, coarse, and pale in colour, but those living in colder regions tend to have longer, shaggier fur. Although it still has rat-like features, the face is deeper and the ears and tail are small; the feet are slightly webbed. They are often heard before they are seen, since they communicate with grunts, squeals and clicking noises. The meat is white and very much like pork.

Tapir Tapirs are found in Thailand, Malaysia, Sumatra, and Central and South America. They are harmless herbivorous animals with pig-like bodies which can grow to over 2m (6.5ft) long, weighing up to 250kg (550 pounds). The thick skin is covered with short, coarse hair. They can usually be found in and around swamp areas, sleeping by day and feeding by night. They are able swimmers and, if pursued, will seek safety in deep water.

Long-Legged Mara The mara lives on the grasslands of Argentina, where it can be seen grazing in small groups, or lying on its stomach basking in the sunshine. In appearance it rather resembles a hare with a large head, a blunt nose and big eyes, long legs and clawed toes. It often grows to 75cm (30ins) long and weighs up to 15kg (33 pounds). It usually moves with a hopping action, but is capable of speeds up to 45kmh (28mph) when disturbed. Maras are easily alarmed; when they run for cover flashes of white fur on the backs of the thighs act as a warning signal. They are burrowing animals, and return to their burrows at night. The entrances are easily identified by piles of droppings.

Wild Pig Wild pigs are common in almost all rain forests, especially in areas adjoining cultivation. Pigs are omnivores; the wild species eat a wide variety of leaves, roots, fruit, reptiles, rodents and carrion. They are usually found living as a family group, although the male will spend much of his time alone. Mature boars have dangerous tusks, and in some areas of South America they are very aggressive, attacking without provocation. Pigs are creatures of habit; if their feeding areas or watering holes can be located they can easily be caught. Although a good food source, most wild pigs are infested with worm.

Ants and Termites Placing a tin can containing a small bait of rotting fruit directly in the path of the ants for several minutes will produce enough protein to last a human for a whole day. Give the can a hard bang on a firm surface, and place it on the fire for about 30 seconds; this will shake the ants to the bottom where the heat will kill them. Then put the can near enough to the heat that the contents will dry but not burn (see Preparing Insects). Worms, beetles, wasps, grubs and very small fish can all be cooked in the same way.

Snakes Ground snakes do not move very quickly and can easily be killed with a stone or stick. Obviously, take care to avoid snakebite at all times – though normally shy, they will strike in self-defence. Having caught your snake, remove the head to make sure it is dead. Slit the stomach skin downward from the neck and peel it back until fully removed. Most of the organs are held in the inner stomach and are easily separated from the body. Snakes make good eating; the flesh is best boiled or fried.

The Preparation of Animals

Skinning and dressing an animal carcass will be much easier if it is done as soon after death as possible. First the carcass should be bled. Smaller and medium-sized animals can be hung upside down from a frame, with the ropes attached around the hocks. The throat should be cut and the blood collected in a container below. Do not throw away this blood; it contains many valuable vitamins, minerals and salt, and once it has been boiled thoroughly it can be used as a food source. It is ideal for thickening and adding flavour to soup.

Note: If you catch an extremely large animal such as a moose or bear, which is impossible to haul up for butchering, you should consider moving your camp to the beast rather than trying to carry it back piecemeal.

Preparing Rabbits Rabbits and small members of the cat family can provide a survivor with a relatively easy-to-catch source of meat. However, rabbits lack the fats and vitamins needed to sustain a survivor's health. Be aware that although a rabbit-rich diet may be easy and tasty, it can also lead to severe malnutrition over a period of time.

When skinning a rabbit, first make a cut behind the head and make sure that it is large enough to

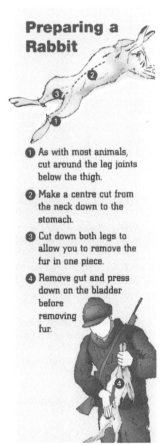

Preparing a Rabbit

1. As with most animals, cut around the leg joints below the thigh.

2. Make a centre cut from the neck down to the stomach.

3. Cut down both legs to allow you to remove the fur in one piece.

4. Remove gut and press down on the bladder before removing fur.

insert two fingers. Peel the skin back and cut off the head and lower limbs. To gut the carcass, cut a line down the belly and open out the body. Most of the innards should fall out when you give the carcass a sharp shake, but make sure that any remaining pieces are scraped out with a knife and washed away with fresh water.

Preparing Rodents Rats and mice are not only edible; they are delicious when stewed with vegetables. Skin, gut and wash them in the usual way; but boil them for about ten minutes before any other form of cooking, to destroy any parasites or bacteria they may be harbouring.

Preparing Hedgehog, Porcupines etc Animals that have a protective coat of spines or a thick shell are best rolled in a thick layer of clay and cooked in the embers of an open fire. Make sure that these animals, especially large ones, are cooked all the way through; any sign of blood when you open it means it is not cooked properly. Crack and remove the clay; the spiny skin will pull off with it to reveal the flesh. Eat only those parts you are familiar with, keeping the rest for bait.

Preparing Insects To humans, insects are not the most appetizing food source; yet any survivor would be foolish to overlook their potential. They are the most plentiful life form on earth, and pound for pound, provide twice the amount of protein as steak.

Insects live both above and below ground; in either case their nests are easily found. Rotting logs provide homes for grubs, termites and beetles. Large flat stones make good nesting sites for a whole host of different species. Remember that insect larvae are also edible and highly nutritious.

Almost all insects are found in abundance, so their small individual size is of little consequence – the mass will provide enough protein. The appearance of insects is also of little importance other than providing the means to recognize its suitability for eating. The secret of dealing with insects lies in how they are prepared.

Insects to Avoid

The following should not be considered as potential food under any circumstances:

• Those with bright colours – all over, or in spots, stripes or patterns. These are usually so coloured to warn animal predators of their poisonous nature.

• Creatures with a hairy skin – again, these may be poisonous, or have stinging contact defence hairs.

• Ticks, flies, lice and mosquitoes, all of which carry disease.

• Those hard-shell insects which may carry parasites.

Monkeys

- Monkeys are a plentiful jungle food resource, relatively easy to trap.

- Native peoples routinely hunt and eat them.

- Western people often feel repugnance at killing and eating monkeys, perhaps due to their superficial resemblance to human babies.

- In survival situations you will indulge this sentimentality at high cost to yourself and your companions. Follow the practical example of the people of the forest. Worry about it after you have reached safety.

This is best done by collecting as many as possible – a minimum of several cupped handfuls. These should be placed in a metal container which has been preheated over a hot fire (a lid of some sort will stop the more active species from crawling out).

It will take several minutes for them to cook, and it is best to turn and shake the container in order to toss the insects and prevent them from burning. Once all the insects are inert, leave them to dry further beside the fire. A good test is to pick an insect from the container and crush it between your fingers; the whole body should disintegrate to a dark brown dust. Next, grind the cooked insects using a stick as a pestle. When this is done pour the powder into a container of warm water; this will separate any unpulverised wings and legs, which will float on the surface where they can be removed. The remaining liquid is little more than a tasteless protein soup, to which edible plant parts can be added to make a nourishing meal.

Butchering Larger Animals

Once your deer, pig, wolf, etc has 'bled out', the carcass can be skinned:

❶ Make the first cut around the knee and elbow joints. Carefully make a full circular cut around the genital organs. Then, starting at each knee, cut the skin down to the abdomen, forming a V-shaped cut.

❷ Continue cutting down the front of the animal, stopping at its neck. Be careful not to pierce the abdominal wall beneath, as

this will spoil the skin. To protect the abdominal wall from the knife, place your hand behind the cut, inside the carcass.

③ Make two more cuts from the front elbow joints in towards the belly.

④ Return to the hind legs and peel back the skin; a cutting and pulling action is best. Continue until the skin has been completely removed.

⑤ Cut open the abdominal membrane – without piercing the stomach or other organs – down to the chest bone. Use wooden skewers to pin back the flaps. Much of the gut will fall from the stomach and drop onto the ground.

⑥ Check that you have removed all internal organs, starting with the windpipe and moving upwards. To clear the entire mass, use a knife to make a deep circular sweep around the genital organs; avoid cutting the bladder.

⑦ After inspection for any signs of disease, keep back the parts of the offal which will be useful (e.g. the kidneys, liver, heart, and the fat surrounding the intestines). Use the rest of the innards for bait, or to make sewing gut. Also keep back the meaty parts of the skull; the brain, eyes and tongue are all edible.

⑧ Once you have cleaned and prepared the meat the skin can be cleaned and dried in order to preserve it.

Fishing

Of all the aquatic foods, fish are the easiest to catch and offer the most obvious form of nourishment. Even with the crudest of fishing equipment, as long as you have knowledge and patience you will be able to catch enough fish for your needs. As with most things, catching fish is a skill and requires practice, and it is unlikely that you will catch much on your first attempt. With growing experience, patience, and the ability to vary your methods according to the situation, you will find a fishing technique which will achieve the results you want.

There are not many general rules that apply to fishing, as they can be caught by a variety of methods – hooks, nets, traps, snares, spears, stunning, poison, and even by simply using bare hands to grab them. All species differ in their feeding habits; however, it is generally accepted that most fish will take bait at dawn and dusk – look for the signs of feeding at those hours. Big fish are hard to catch as they are heavy and full of fight; if you do not have the correct fishing kit in your survival pack, then improvise. If you see large fish close to the surface try using a spear or bow and arrow, stalking your fish with extreme care in order to get close enough.

When and Where to Fish

In hot weather fish will tend to seek cooler water, either in deeper river pools or under shade; these are the places in which to cast for fish. The outer bank of a river bend also holds deeper water and this may be a good place to fish, especially if water levels are running low. Deep lakes are also good bets in hot weather. Fish tend to shelter below underwater rocks and logs or undercut riverbanks.

At dawn and dusk, fish tend to prefer shallower water or can be found around the edges of a lake or pond. Fish need a certain amount of warmth and will seek out warmer water. They also tend to feed better in shallow water. Fish will always lie in the water facing the oncoming current. This enables them to spot any food coming towards them, and also ensures a better flow of water over their gills. Knowing this, you will have better success if you let your natural bait move downstream towards likely shelter spots at a natural pace, so that they can see it and hopefully accept it as a normal piece of food.

Fish also like to be where the water is well aerated, such as at the bottom of a small waterfall. When using natural bait, cast it into the cascading water and let it move naturally down and across the pool, or for a little distance downstream if in moving water. Then, very smoothly and quietly, bring the line in and cast again as

Look for Fish:

In warm weather:
- In deep pools and lakes.
- Under shade, and undercut banks.
- On the outside of bends.

In cool weather, and at dawn and dusk:
- In shallows.
- Round the edges of ponds and lakes.

In any weather:
- Under white water.

Fish Hooks

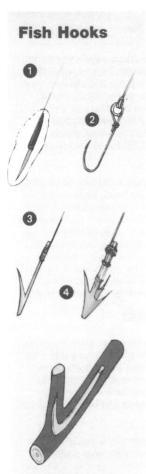

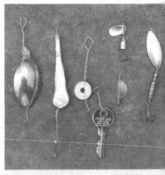

Commercial and improvised lures.

Improvising a fish hook can be simpler than it seems.

1 A simple bait covered gore.
2 Stiff wire or safety pin.
3 Whittled hardwood.
4 Strong dog rose or similar thorn.

before. The best pole to use for this type of fishing is a slender, flexible one, as this enables the line to be pulled gently out of the water instead of being dragged back through it. This type of pole also makes casting and recovery a lot less effort for the angler. If you are fishing for carp, catfish or eels you need to be aware that they feed on the muddy beds of slow-moving rivers and ponds. With this in mind, bait will need to be cast on the bottom and then moved very slowly.

Bait Your first choice for bait should be food that is normal to the fish's diet. Before you start to fish on a stretch of water, study it and the surrounding shore for morsels normal to the fish. Look for insects, worms, shrimps, minnow or shellfish. If none of this natural bait is available you will have to substitute an alternative, such as small scraps of meat or artificial substitutes. Fish are often attracted by the struggles of live bait. Try using a grasshopper or a beetle and see if it is taken by a fish. If it is, take another insect and carefully impale it on the hook without killing it. This should attract another bite from the fish, which this time will end up being caught. Minnows can also be used as bait in this manner, but under the water. The hook should pass through the body under the backbone and to the rear of the minnow. A float will be needed to keep the bait off the bottom of the water.

Lures and Hooks A lure is some form of artificial bait. It is designed to look like an insect or a small fish in order to fool the fish into thinking that what it sees is its natural food. A convincing

appearance alone is not enough; the angler must also be able to manipulate the lure in order to mimic the movements of live bait struggling in the water.

Lures can be improvised from many sources of material. They can even be made from a tuft of hair (from your own head if necessary), feathers, a scrap of brightly coloured cloth, or a fish fin with a piece of flesh attached. In fact, anything will do as long as it looks like an insect of some description. The lure should be constructed around the hook so that this is hidden.

Your basic survival fishing kit should contain a good supply of variously sized hooks. Good fishing hooks can also be improvised from a wide variety of materials and items – thorns, safety pins, wire, etc. Always make sure that your hook is the correct size for the fish you are trying to catch; and that, once you get a bite, the hook will stay attached to the line.

Fishing Hints

Fish tend to be very wary, and will swim away and hide at the first sign of anything they perceive not to be in their normal pattern of events. They are able to detect even the slightest vibration in the water, and are even aware of heavy footfalls on the bank. Therefore it is vital that when you approach the edge of the water you do so slowly and gently so keep any ground vibration to a minimum. Keep as low as you can, as quiet as you can, and move as little as possible. Never let your shadow fall onto the water.

Put your bait into the water slightly upstream from the location of the fish, and allow it to drift downstream with the current until it has passed you. If by that time no fish has taken the bait, gently recover it and try again. If no fish take the bait after a few tries, change your fishing pitch – but remember to make your move slowly and quietly. If you still are having no luck, try again at the opposite end of the day, or even after dark if the water is clear and shallow.

Attracting Fish Attracting fish to a feeding ground is a good way to ensure a bite. Many anglers throw ground bait into the water to lure fish into what they perceive to be a good feeding ground. If you have plenty of bait it is recommended that you do the same. An alternative method when fishing in a pond, pool or lake is to tie a piece of unwanted offal or carrion to a branch overhanging the water. This will attract blowflies to lay their eggs in the

Night Line

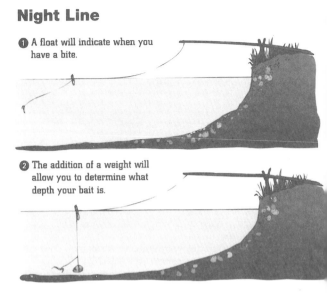

1 A float will indicate when you have a bite.

2 The addition of a weight will allow you to determine what depth your bait is.

meat. After a few days maggots will appear, and will fall into the water at a steady rate, thus attracting the fish. If you then place one on your hook you should soon catch something; better still, use a large net.

Night Line Fishing   A night line consists of a line with one or more hooks which is left in the water all night. The hooks (prefer-ably gorge hooks) should be baited with something that cannot easily be lifted off by eels, such as a small fish or a small piece of meat. The line should then be fastened firmly to a rock or a stake on the bank or an overhanging branch. The line should be checked every morning for a catch; if there is one, remove it and replace the bait. The line can have a single hook, or several stretched on a line across the river. The depth of the hook can be adjusted with weights to catch a variety of fish under most conditions.

Fish Traps   Fish traps can be used in both fresh and sea water. The type of trap required will be dependent on the water in which you are fishing and the size of the fish. Traps can be made to be portable or permanent. The most common form of fish trap is the portable basket type, built with a cone-shaped entrance. They can be constructed from hazel or willow sticks, reeds or bamboo, or improvised from man-made discarded containers such as a plas-tic bottle. This shape makes it easy for the fish to get in and almost impossible to escape. Once constructed, the pot should be baited and placed facing upstream in a river or a rock pool.

Permanent Traps   An on-site trap can be made by piling stones or driving wooden stakes into the riverbed to form a pen. It may be necessary to form the trap in such a way that fish will be funnelled through the entrance into a secure compound. The sit-ing of the trap is critical, but where possible full advantage should be taken of natural features which will enhance your catch and save time and energy. It may be possible to herd fish by wading into the water starting 100 metres upstream and walking towards the trap. Trapped fish can be speared with a sharp stick.

Fish and Wild Fowl Snares Snaring fish is not as easy as snaring small game, but it works on similar principles and can be achieved with time, care and observation. Take a normal animal snare and attach it to the end of a stick, which should be at least 200cm (80ins) long. Tie an extension on to the end of the snare wire so that you can close the noose at will. (see page 95.)

Fishing Nets It may even be possible for the survivor to make and use his own fishing net – a gill net – as long as he has enough line (a possible source for this would be a parachute rigging line). First decide how long the net has to be, and then tie the top line of the net between two saplings or stakes the right distance apart. Tie another line below this to define the width of the net. Any nylon lines that are available should have their inner cores stripped out. Take a piece of this inner core and double it. Tie it to the upper of the two lines using an overhand loop. The two ends, which should be 30% longer than the required width of the net, should then be allowed to hang down loose. It is now just a matter of repetitive knotting; but you probably have lots of time on your hands. Depending on your location, the materials available and your ability to manage it, make your net as large as possible.

Providing a stream is not too wide nets can be erected right across it; if it is, then the stream can often be dammed to make it narrower. Support the net by stretching a line across and secure the bottom edge in the water with heavy stones. In a larger body of water, such as a river, nets should be set just above or below an eddy.

Wherever they are set, it should always be in a stretch of quiet water. On a lake shore the net should be set at right angles to the bank, preferably off a small headland.

Minnow Traps Small fish such as minnows can be found in most water, especially where it is shallow, such as at the edge of a river or a lake. A normal net will be too large, but several fine-meshed nets can be constructed from a pair of women's tights. Cut a length off the tights, knotting one end if it is not the toe

Portable Fish trap

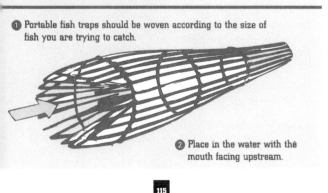

① Portable fish traps should be woven according to the size of fish you are trying to catch.

② Place in the water with the mouth facing upstream.

Fish Pen

1 Fish funnelled through entrance to pen.

2 Where possible, use natural features such as rock pools to trap fish.

3 Once in, the fish may be netted or speared.

piece; splay the open end around a ring of stiff wire, and secure this to a forked branch. A well-perforated tin can attached in the same way will serve the same task. Do not disregard the food value of very small fish; dried and roasted in quantity, they make good eating.

Eel Traps A simple eel trap can be made from a suitable box. On each side of the box, near the top, make a couple of small holes. Inside the box lay some ripe meat as bait. Weight the box and put it into the water, checking it every two or three days. (To do this, take the box out of the water first and part-empty it before opening it to see what you have caught. Eels are proverbially slippery, and it is almost impossible to hold on to them long enough to lift them out of the water.)

Water can be channelled into natural features to trap fish.

Spear Fishing

- Stand directly above a 'run' used by fish.

- Use a strong, three-pointed spear.

- Don't throw it – stab directly downwards.

- Chase any wounded fish.

Spearing Spearing fish can be difficult, especially in the hours of daylight, but with practice it is possible. The fish will need to be fairly plentiful if you use a single-point spear, and you are advised to use a three-pronged head or trident for greater efficiency. The

Making a Fish Net

① Place your top line between two trees and hitch a double length of cord every 3cm.

② Cross tie the hanging strands with a simple granny knot to form a diamond pattern.

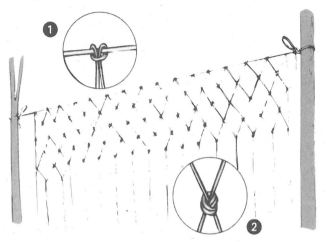

Providing you have the materials, making a fish net is always a good idea. Not only can it be used for catching fish and game, it is also handy when gathering plants.

spear should be stabbed into the water at the fish, not thrown; and the stab should be aimed directly downwards. Spearing in this way reduces refraction in the water, and thus the risk of misjudging the angle; and will also pin any speared fish to the streambed. The best position for this method is standing directly over a fish run. Make sure the spear is strong enough to withstand the thrashing of a large fish. Always chase after a badly wounded fish, as they will not go far.

In much the same way, fish can be shot using a bow and arrow. Use long arrows which will be visible above the surface and will restrict the movement of any wounded fish trying to escape.

Killing or Stunning with Poison Poisons, usually derived from plant sources, are used by some native peoples to catch fish. Fish are quickly affected by poison being introduced into the water, normally rising to the surface quite soon. The speed at which the poison works will depend on the water temperature. Around 21

Fishing hints

• Approach the fishing water making as little noise and vibration as you can.

• Don't let your shadow fall on the water.

• Make no sudden movements or noises.

• Try to spot the fishes' natural prey in that water, and bait with it.

• Spread ground bait.

• Cast up-stream and let your bait float down to the fish.

Cleaning Fish

• As soon as you land a fish, kill it by hitting the head with a stone or club.

• Slit the stomach from jaw to tail.

• Scrape out the innards.

• Wash out the body cavity thoroughly.

• Hold the tail and scrape off the scales, moving down towards the head.

• Unless you are going to cook on a spit, cut off the head.

• Don't eat fish raw – cook by spit-roasting, baking in embers, or frying.

• Keep head, tail and guts for bait.

Crustaceans

• All crabs, crayfish, shrimps, prawns, etc, start to spoil the moment they are caught.

• Cook them at once, by dropping them alive into boiling water.

• Only thorough cooking destroys any organisms they may contain.

• Throw away the innards and the gills.

• The meat inside the shell and claws is edible.

• Failure to cook quickly enough, or thoroughly, invites serious food poisoning.

degrees C (70 °F) or warmer is ideal.

One of the most common poisons is rotenone, a substance found in tropical plants that stuns or kills cold-blooded animals while leaving the flesh safe to eat. The following plants can be used to stun or kill fish:

- **Derris eliptica** is the main source of commercially produced rotenone, which is a natural pesticide. The roots from this large order of tropical shrubs and woody vines are ground and mixed with water, which is then thrown into the river. Where possible the mixture is best left overnight to strengthen.
- **Anamirta cocculus** is a woody vine which grows in southern Asia and on islands in the South Pacific. Crush the bean-shaped seeds and throw them in the water.
- **Croton tiglium** is a shrub or small tree that grows in waste areas on islands of the South Pacific and which produces seed capsules. Crush the seeds and throw them into the water.
- **Barringtonia** is a large tree that grows near the sea in Malaysia and other tropical regions. It produces a fleshy one-seeded fruit that can often be found rotting on the ground. Both the seeds and the bark of this tree can be used as fish poison by crushing and throwing into the water.
- On the coastline, burning coral or seashells can produce lime. The white dust residue can be thrown into the water.

Crustaceans Crustaceans include crabs, lobsters, crayfish, shrimp and prawns. All are edible, and can be found in fresh and salt water around the world. Most are best caught at night,

Gutting and cleaning fish before cooking; split open the stomach (top), scrape out the inside (below).

using a light such as a torch held near the surface of the water. Mussels, limpets, clams and periwinkles can also be eaten, as can scallops, sea urchins and starfish. To clean a crustacean, throw away the intestines and gills – the rest of the meat, including that inside the shell and claws, can be eaten.
Warning: All crustaceans must be thoroughly cooked as soon as they

Clean thoroughly in fresh water before cooking.

The Dakota Hole

A Dakota fire is not easy to construct but essential if you wish to keep your presence unknown.

are caught, as they do not keep. If you delay in eating any crustacean or you fail to cook it properly, then you run the risk of the worst type of food poisoning.

Frogs Small amphibians such as newts and frogs can also provide a good meal. They are to be found around fresh water, usually revealing their presence by croaking. However, any croaking will stop as you approach, so have plenty of patience and keep still until they are fooled into thinking that you have gone away again. During the mating season catching frogs becomes quite easy. All you need to do is to splash the back of your hand gently against the surface water and frogs will jump onto it. It is possible to catch several in as many minutes.

Preparing Fish, Snakes and Amphibians Once you have caught a fish you will need to bleed it immediately. To gut the fish, slit open its stomach from the lower jaw to the tail and scrape out the innards. Wash the area thoroughly to flush out any remaining pieces. Fish can be cooked with their scales on, but if you have the time they make more pleasant eating with their scales removed. To do this, scrape downwards with a knife from the tail to the head. Fish such as catfish and sturgeon, which do not have any scales, can be skinned instead. Smaller fish, e.g. those less than 3 inches long, do not need gutting, but some will still need scaling or skinning. The head should also be cut off unless you are going to cook the fish on a spit. Raw fish may contain parasites, and should only be eaten cold if the means to cook are outside your capabilities. This should very seldom be the case if you have a fire; fish can be cooked in a wide variety of ways – spit-roasted, baked, boiled or fried. Fish heads, tails and intestines all make good bait.

Snakes and Reptiles Skin a snake by cutting off its head and slitting its body skin from the severed end downwards for about 20cm (8 inches). Peel back the skin to the length of the cut; grip the flesh and continue pulling the skin downwards until within a few centimetres of the tail; then cut off the remainder. If the snake body looks bloated or lumpy, split it open and remove the innards. Cut the body flesh into small sections and cook – roasting or boiling is best.

Lizards, frogs and turtles are good to eat. Before cooking, take off the head and skin; this is particularly important in the case of frogs, as their skins may contain a poison. Turtles will need to be boiled first to remove the shell. The turtle meat can then be sliced up and used to make a tasty soup with vegetables.

Molluscs Shellfish make an excellent base for a soup to which vegetables can be added. In addition, they can be boiled, steamed or baked in their shells.

Methods of Cooking

The proper preparation and cooking of food will make it safer to eat as well as more appetizing and digestible.

If possible you must try to have one hot meal a day. Most foods, whether animal or plant, require some form of preparation, whether washing, cleaning, scaling, plucking or skinning. How you cook the food also makes a difference, and will eliminate wastage. In the event of a food surplus the survivor is advised to prolong its edible life by preserving.

Roasting Stick Initial roasting should be done over a high heat, which will crust the outside of the meat and seal the juices in. This is followed by slowly turning the meat over a more placid flame. The dripping juices will cause the flames to flare and burn the meat; prevent this by placing a tray below the roast. These juices

can be used to baste the roast, and improve its flavour. Larger animals (those larger than a domestic cat) should be cut into small pieces before roasting. These can be roasted by simply pushing the meat onto a stick and holding it over or near hot embers. If you do not want to sit and hold the stick you could construct an arm or a crane.

Automatic Spit With a little ingenuity you can prevent food from burning by constructing a spit which is turned by the wind. This not only cooks the food evenly, but saves the time which is otherwise wasted while watching to make sure the food does not burn. A normal crane is fitted with a wire line (cord will burn through) on which the meat is hung. Adding a flat slab of bark about the size of your open hand will

A roasting stick.

Attaching a 'flag' to your Billy can will cause the breeze to gently turn it thus cooking your food evenly.

allow the wind to twist the meat; a natural counter-action will turn it back the other way.

Boiling Tough meat will need to be boiled to tenderize it, even if you intend to finish it off by some other cooking method. Any nutritional value leached out of the meat through the cooking process will also be retained in the water, which makes boiling a very efficient cooking method as long as you retain and use the water. One thing to remember, however, is that the higher in altitude you go, the longer it will take for water to reach boiling point due to the reduction in air pressure. Above 4000m (12,000ft) cooking raw food by boiling becomes almost impossible and should not be attempted.

Water can be boiled even if you have no metal container – you just have to look for alternatives. Suitable containers, which will not burn while the water is boiling, include half a green coconut; or a short length of bamboo, cut just below each of two joints; or a suitable large shell. A large single banana leaf, or a waterproof container made from a piece of folded birch bark, will also hold water while it is boiled as long as the fire is kept low and the leaf or bark kept moist. Use thorns to pin the folded corner points.

Baking Baking is a less intense form of cooking than roasting, and the heat is more constant. To be baked food must be enclosed, either in an oven, or in a wrapping of leaves or clay in a pit under the fire, or in any closed container. Baking is best done with glowing coals rather than flame.

Steaming Steaming food can be done without having a container. It is best for foods which do not need much cooking, such as tender greens. The other advantage of steaming is that it retains more of the nutrients in the food than any other method of cooking. To use this method, first dig a pit and place in it a layer of heated stones. Cover the stones with leaves and put the food to be cooked on top of the leaves. Use more leaves to cover over the food, and push a stick down into the food space. Finally, pack a top layer of earth over the leaves and around the stick. Withdrawing the stick will leave a little hole that leads down to the food space.

When water is poured down this hole it will hit the hot rocks and turn to steam, slowly but effectively cooking your food.

Hangi The hangi is a cooking method which originally came from Polynesia. It is slow and safe, especially if you need to be away from camp, or if you have no utensils to cook your food in. First, dig a pit five times the size of the food parcel it must accommodate.

The Hangi

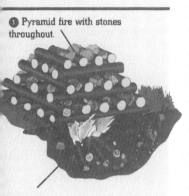

① Pyramid fire with stones throughout.

② Fire pit five times larger than food parcels.

③ Covering of foliage and fresh earth.

④ Food parcels wrapped in large leaves.

A well-prepared hangi will cook your food while you are away from camp.

Lay your tinder and kindling in the middle of the pit. Make a pyramid fire above the hole, placing each layer of logs at right angles to the last layer. Build this up to a height of about six layers, with fist-sized stones and rocks placed between the layers. (Do not use soft, porous or flaking stones such as limestone, as these may explode when exposed to heat.) Once the fire is established it will burn its way through the log pyramid and the hot stones will fall into the pit.

Rake the hot embers to the side of the pit exposing the hot stones. The food should be wrapped in large, clean leaves (make sure that they are not poisonous) and placed on the stones. Meat and any other food which needs the most cooking should be placed at the hottest point of the pit, that is, the centre. Softer foods such as vegetables should be placed nearer the edges. Once all the food has been packed into the hangi, cover the pit with a roof of foliage and seal it with the earth spoil, to keep the heat in and prevent animals from foraging. The food may take three to four hours to cook in this way, but the advantage is that it will not become overcooked even if left for up to eight hours.

Improvised Haybox Another method of slow-cooking food is to construct an improvised haybox. This will prove especially valuable where firewood is scarce, and allows for a meal to cook safely while you are out foraging or attending to more pressing matters. Another benefit of the haybox is that it cooks food well and cannot overcook or burn it.

A box or container is lined with a thick layer of insulating material; if no such container is available then a polythene bag will do. As the name implies, hay was used as the insulating material but more modern insulating materials such as polystyrene or crumpled newspaper can also be used. The other requirement is a can which will act as a cooking pot, preferably one with a well-fitting lid. Heat your meal over a fire until it starts to boil, then seal the cooking pot with a tight-fitting lid. Place it at the centre of your haybox and surround it with well-packed insulating material. Leave for approximately five hours before opening. It is advisable (although not necessary) to bring the pot back to the boil over an open fire before eating.

Food Preservation

Food-gathering may not always be successful: there will be times when game will be difficult to find or catch. The survivor must not rely on the assumption that a regular supply of wild food will always be available. In these circumstances, knowing how to preserve and store foods is a valid survival skill. Preserved food will not only back up fresh supplies but may also be carried with you if you plan to move on.

The aim of food preservation is to prevent the deterioration of the food and so prevent wastage. Meat can be dried either in the sun and wind or else over a fire. The aim of drying is to drive as much of the moisture content from the meat as possible. This not only concentrates its nutritional value, but will also preserve it longer from decomposition and moulds. A piece of dried meat should contain only about 5% of its pre-dried content of moisture. Meat should be cut into long, thin strips and placed to dry on a platform safe from scavengers but open to the sun and wind. The

Air-dried meat.

process may take up to two weeks, and during this time the strips need to be kept dry from any rain and free from flies.

In warm or damp weather when meat deteriorates rapidly, smoking over a low fire can save it from spoiling for some time. Care must be taken to keep the meat from getting too hot. Cutting it across the grain into thin strips and either drying it in the wind or smokng it will produce 'jerky'. Fish should be flattened by removing the backbone, and skewered in that position for smoking.

A small version of a North American tepee with a platform constructed in the middle makes an excellent smokehouse. By tying meat to the upper ends of the poles and closing the smoke flaps a good concentration of smoke is obtained. Try to create a fire with little flame which produces quantities of smoke. The meat will be ready when it is brittle.

Plants, leaves and fruits can also be dried by the methods described. To dry fruits successfully, cut them into thin slices first. Berries are best preserved by being turned into jam or jelly.

Keeping Meat

• When you have more meat than you can eat in a day or two, preserve it for leaner times or for travelling.

• If near the seashore, boil and distil seawater to extract salt. Rub salt into meat and air-dry it; or store in a salt-heavy brine solution.

• If inland, in a hot climate, slice meat thinly and dry in the sun.

• If the sunlight is inadequate, smoke thin slices slowly over a fire.

• In a cold climate, cut meat into small strips and freeze it.

Navigation in the jungle is difficult even for the most experienced as heavy vegetation and irregular terrain make travelling in a straight line impossible. A compass must always be used but natural features may be easier to follow.

Follow a ridge rather than hacking through endless stands of bamboo and clinging vegetation. River courses meander, sometimes increasing the linear distance three-fold; however, travelling down a river by raft is a good idea if it is feasible.

Tracks Many tracks can be found in the jungle, the smaller ones made by animals. Those that are well worn normally lead to water – if you find one, follow it downhill to water. Man-made tracks may either link villages or be habitual hunting routes. Logging tracks may also be found, driven great distances into the jungle in search of particular types of tree.

Maps A map is a sheet of paper on which an aerial view of the area it represents is drawn. The detail included on the map will vary, depending on who did the survey, what use the map is intended for, and the scale.

Most maps have one thing in common: almost all have a grid overlay dividing the map into squares which are either lettered, numbered or both. Most maps also contain a legend explaining the scale, distance and symbolized features.

Different maps are made to suit different purposes. Aircraft operators and strategic planners normally use maps with a scale of 1/250,000; these provide only generalized information and show only principal features. Maps used by the military for route selection are normally to a scale of 1/50,000; these show detailed features and relief. This book uses a 1/50,000 scale map to illustrate the meaning of elevation, contour intervals, conventional signs, the grid system, and information on magnetic variation.

Compasses A compass is a precision instrument which allows the user to identify North and thus the other cardinal points. Used in conjunction with a map, it allows for navigation over the surface of the planet. There are many shapes and sizes of compass, but

LIFESAVER
The Compass

This compass is constructed with a clear plastic base and a compass housing which contains the magnetic needle. The base of the compass has a magnifying glass and is etched with a variety of scales and a number of romers – scales to accurately divide a grid square into tenths to help calculate grid references. The rim of the compass housing can be rotated, and is marked with segments showing degrees, mills or both, while printed on the base is an arrow and orienteering lines.

The 'bearing' gives the direction to a certain point. It can be defined as the number of degrees in an angle measured clockwise from a fixed northern gridline ('easting'). The bearing for North is always 0/360 (the number of degrees in a circle) or 0/6400 (the number of mills in a full circle).

Map Scale

Maps to a scale of 1/50,000 show all roads, tracks, paths, rivers, streams, lakes and most man-made features. In addition they indicate ground relief using contour lines, and areas of forest. The numbered grid divides the map into 1000m (1,093 yard) squares that convert into longitude and latitude bearings.

Finding Your Way

• Given a rough idea of your position relative to inhabited areas, you can find and follow an approximate direction across country even if you are a lifelong city-dweller.

• Map, compass, and basic knowledge of their use are very valuable; but even without them you need not be 'lost'.

• You may have the means to make yourself a simple compass.

• Even if not, the sun by day and the stars by night will always show you the direction of North.

all work on the principle of a magnetized needle continually pointing North. Always remember that any compass works on the magnetic attraction situated close to the North Pole; local power supplies or heavy metal objects can pull the needle from its correct course. Most compass manufacturers dampen the movement of the needle by filling the compass housing with a liquid. This sometimes produces a bubble, but providing that this is not large it should not affect the operation of the compass. This book uses a 'Silva'-type compass to illustrate the various uses, and its relationship to a map.

Conventional Signs Every map has a panel of conventional signs which indicate a variety of objects such as roads, railways, rivers, cliffs, buildings, etc. In a survival situation identifying a man-made object on a map which covers your area can lead you to safety. A complete study should be done of any map in your possession to establish not only your position, but also what else is in the area. Even if it is barren of any human structures it may guide you to a more habitable place of food and shelter.

Contours Contours are used to represent different elevations – e.g. valleys, hills and mountains – on a flat surface such as a paper map. They are intended to give a perspective view of shape and elevation. Each contour line follows the same height around the hills, into the re-entrants, and over the spurs. On the 1/50,000 Ordnance Survey map the contour lines are 50 feet apart. Therefore, if the lines are close together it follows that the land is rising very quickly, and if far apart, that the slope is gentle. Contour values, which are given to the nearest metre, are marked so that they read facing uphill. Remember, however, that while the heights are in metres, the contour lines are 50 feet apart.

Setting a Map by Inspection If you possess a map that you know to represent the local area, you will be able to orientate yourself. Look for an obvious and permanent landmark, for example a river, road or mountain. Identify the feature on the map, and

Reading Contours

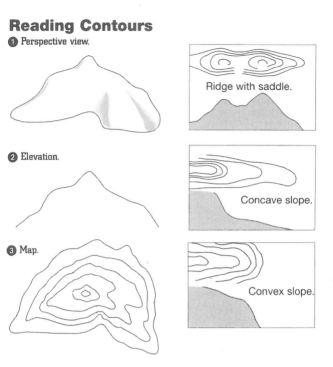

1 Perspective view.

Ridge with saddle.

2 Elevation.

Concave slope.

3 Map.

Convex slope.

then simply align the map to the landmark. The map is now 'set' to conform to the surrounding features. If you do not have a compass, make a note of a distant feature at one of the cardinal points – North, East, West or South.

Setting a Map by Compass You can set any scale map and align it with the surrounding terrain by simply using a compass. Choose a North-South grid line on the map, and lay a flat edge of the compass along it with the direction arrow pointing towards North (top of the map). Then, holding the map and compass together, turn both until the compass needle points North. The map is now set to conform to the surrounding features.

Finding a Grid Reference Almost all maps are covered with light horizontal and vertical lines, each marked with a two-digit number. These are called grid lines – on a 1/50,000 map they are blue and spaced 1km (0.62 miles) apart. The vertical lines are called 'eastings': these are always given first when quoting a grid reference. The horizontal lines are called 'northings'; these are given after the eastings. The numbers straddling the left grid line of the easting and the centre bottom of the northing defines each grid square – an example is illustrated on page 130.

As each grid square measures 1000m by 1000m – which is quite a large area – it is desirable to reduce this by calculating a six-figure grid reference. This is done by mentally dividing the square into tenths – e.g. halfway up or across a square would be '5'. This reference is then added after the relevant easting or northing figure. To gauge the tenths accurately use the romer on the compass, or a protractor as illustrated.

Grid Reference

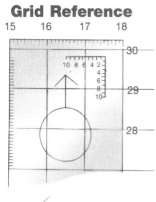

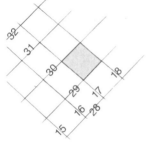

A compass with a built in romer makes it easy to subdivide a grid square and produce a more accurate six figure reference.

The example shown is 17528

Taking a Compass Bearing from the Map Once you have established where you are, and where you wish to go, you can work out your route. Plot the most logical route to your objective, taking into account the distance, terrain and any obstacles. Divide your route up into 'legs', if possible finishing each leg close to a prominent feature, e.g. a bend in a river or the corner of a forest area. If using a 1/50,000 scale map the ideal distance for each leg should not be longer than the length of your compass base.

First take a bearing from where you are (call this point A) to the feature at the end of your first leg (call this point B). Place one edge of the compass along the line joining A to B, making sure that the direction-of-travel arrow is pointing in the way you want to go. Hold the compass plate firmly in position and rotate the compass dial so that the lines engraved in the dial base are parallel to the North-South grid lines on the map.

Finally, read off the bearing next to the line-of-march arrow on the compass housing. To walk on this bearing, simply keep the magnetic arrow pointing North over the etched arrow in the base, and follow the line-of-march arrow as illustrated. Repeat this process for each leg.

Magnetic Variation When we talk about 'North', bear in mind that there are three Norths. True North is not generally used in navigation; it is the fixed location of the North Pole. Grid North is more familiar – it is the North indicated by the gridlines on a map. Magnetic North is where the needle on the compass always points due to the strong magnetic attraction generated by the Earth's magnetic field. However, the direction of magnetic North may vary by a small fraction from year to year due to changes in this magnetic field. This difference can be calculated using the information shown on the map, i.e. the date it was printed and the degree of variation. This variation is then either added or subtracted to Grid North to get a more accurate bearing. Put simply: 'Mag to Grid, get rid' – i.e. subtract the variation from your compass bearing before applying it to the map. 'Grid to Mag, add' – i.e. add the variation to your map bearing before applying it to your compass.

Keeping on Course Three factors will determine which route you take: the weather, the time of day, and the terrain between you and your final destination. In good visibility, select features that are both prominent on your map and visible to the eye. Once you have taken a bearing, choose a feature along the line of march and head towards it. This saves you constantly looking at your compass. A back bearing will help keep you on course if the terrain pushes you off track, i.e. you are forced to avoid some obstacle. Mistakes in poor visibility can be avoided if you consult the map every time you meet a prominent feature. Careful study of the map should provide you with a mental picture of the ground relief, which will in turn warn you of any obstacles such as rivers or marshland.

Global Positioning System (GPS) Developed by the United States Department of Defense, the GPS system consists of 24 military satellites which orbit the Earth, continually giving out the time and their position. Receiver units on Earth pick up this information and use it to determine location. Designed for the military, the system now guides most of the world's shipping, aircraft (and 'smart' missiles).

Many outdoor workers and enthusiasts also carry hand-held units, no larger than a mobile phone. Receiving units commercially available to civilians vary, as does their accuracy. A deliberate error, called Selective Availability (SA), was built into the system. This 'dithers' the signals so that only a Coarse Acquisition (CA) can be obtained, therefore reducing

Magnetic Variation: Don't Panic!

Some people put a lot of emphasis on adding or subtracting the magnetic variation, but for survival purposes you can virtually forget about it. By simply shortening the legs of your selected route and choosing a prominent feature to march towards or from, you reduce the risk of magnetic variation error. Some will criticize me for saying this; but I have never bothered to work out the magnetic variation when walking – and I have walked across more desert, jungle and Arctic tundra than most. That said, it is advisable to adjust the variation when plotting long routes across barren land, or when travelling by vehicle.

LIFESAVER
GPS Limitations

The GPS requires tuition in its proper use, and is not a compass in the normal sense. Despite its excellent qualities, the GPS system can be shut down at times. In addition, the units have a high battery drain. Do not become complacent – don't forget your compass.

accuracy to about 40m (130 feet). The SA can be overridden for military use by a 'P' code, giving an accuracy of about 10m (33 feet).

How it Works The GPS receiver searches for and locks on to satellite signals. The more signals you receive, the greater the accuracy, but a minimum of four is required. The information received is then collated into a usable form – for example, a grid reference, height above sea level, or a longitude and latitude.

By measuring your position in relation to the satellites, the receiver is able to calculate your position. This is called satellite ranging. It is also able to update your position, speed and track

GPS

❶ Hand held unit locks on to any satellite in its range.

❷ At least four satellites need to be available. The more contacts, the more accurate will be the information returned.

❸ Satellites return signal to GPS unit

❹ The unit translates the signal into information required; grid reference; height above sea level etc.

while you are on the move; and can pinpoint future waypoints, thereby taking away the need to find recognizable landmarks.

Making a Sketch Map All aircraft and most vehicles which venture away from civilization carry maps, and these should be located before attempting to improvise a sketch map. If you are left with no option, the sketch map will help you to plan any travel, indicate to others the route taken, and provide an easy reference to navigation.

A sketch map is best drawn similar in orientation to that of a normal map, i.e. with North at the top centre of your sketch. Choose a high vantage point to observe the surrounding countryside, and fill in the detail working outwards from the centre. Divide your map using a distance grid ruled along the top and down the left side. Next draw in all prominent features – mountains, rivers, forests, marshes, etc – using the grid as an estimate of distance. If no coloured pens are available, black and white line drawings are fine; information can be qualified by the addition of notes. Mark camp sites and routes as you travel.

Electronic Compass Electronic compasses have been around for a number of years, but until recently they have not behaved well and were fairly unreliable. An improved generation has started to appear which provides something between a magnetic compass and GPS. The electronic compass has a number of features well suited to the survival situation. These include course memory, night vision back-light, automatic route reverse, stored bearings, and a clock which gives time and distance. Electronic compasses are simple to operate, and are extremely handy during the hours of darkness. Power is a consideration, as the unit runs on batteries; however, the battery life is around 200 hours and most are fitted with automatic shut-down.

Back Bearing

If you become disorientated there is a simple way to pinpoint your position and keep yourself on course. This is called a back bearing; as the name implies, it is the opposite to a forward (normal) bearing. For example, if a bearing is 260°, and you subtract half a circle – 180° – you get a back bearing of 80°. If the original bearing is less than 180° you simply add 180°; e.g. if your bearing is 60°, plus 180° gives you a back bearing of 240°.

This has several uses. If you leave a prominent feature and move across rough terrain which forces you off your line of march, you can always double-check by converting your bearing to a back bearing and re- fixing on the feature behind you.

By converting the bearing of two landmarks that you can identify on the map, it is possible to establish your correct position. Take a bearing to the first landmark; e.g. say this is 280°. If you wish, calculate the magnetic variation, which we will say is 5°, and subtract. This leaves us with a revised bearing of 275°. Since this is greater than 180°, a back bearing can be achieved by subtraction of that number, i.e. giving us 95°.

This bearing is applied to the compass dial and the edge of the base is set against the landmark shown on the map. Pivot the whole compass until the orienteering lines in the base of the housing are running parallel to the eastings. This should allow you to draw a line from the feature at an angle of 95°.

Repeat the whole procedure for your second landmark, and draw another line. Your position is indicated where the two lines cross.

Improvised Compass By magnetizing a small, straight piece of metal such as a needle, pin or razor blade, and suspending it so that it can swing freely, it is possible to make a simple compass. The piece of metal can be magnetized by stroking it in one direction with one pole of a magnet. Magnets can be found in any radio set, installed as part of the speaker. The pointer will then need to be suspended – with a small object such as a needle or razor blade this can be done by floating it on water.

To do this, improvise by sticking the pointer through a small piece of cork, or by balancing it on three matchsticks or a small twig. **Warning: Remember that the container which holds the water must not be made of metal, as this may affect the magnetic field and cause the pointer to give a false reading. Also, remember that after a time the magnetizing effect on the pointer will wear off and will have to be repeated.**

Steel pointers can also be magnetized with electricity. For this you will need a battery that is capable of producing more than 6 volts, and a good length of insulated copper wire. Suitable wire may be found in the coils inside radios and generators, or any electrical equipment in most kinds of vehicles. To magnetize the pointer, wrap the wire around it as many times as you can. Strip the ends of the wire of any insulation, and attach to the battery for 15-30 minutes. Ensure that the wire wrapped around the pointer is long enough or it will get too hot. If this happens, disconnect from the battery and allow them to cool before starting again. Once the pointer has been magnetized (this may take several attempts), the end indicating North will be that nearer to the negative battery terminal (remember: N stands for both Negative and North).

Finding Direction Without a Compass

Compasses and GPS systems may be the easiest and most convenient methods of finding a direction, but survival starts in the most unusual places, and the odds are you will be without either. Luckily the most important aspect of survival navigation is direction, and this can be established through a number of time-honoured methods by using a bit of intelligence.

Stick and Stone Method The accuracy of this method depends on using level ground and marking the shadow with some degree of accuracy. A North/South indicator can be produced if a line is drawn at right angles to your East/West line; any other direction is simply a calculation from these cardinal points.

- On a sunny day, find or cut a stick about one metre (39ins) long, and push it upright into some level ground.
- The stick will cast a shadow. Using a small stone, mark the end of the shadow as accurately as possible.
- After 15 to 20 minutes the shadow will have moved. Using a second stone, mark the tip of the new shadow.
- On the earth, draw a straight line running through the positions of both stones. This is your East/West line.
- Put your left foot close to the first stone, and your right foot close to the second. You are now facing North.

Using a Watch Using an analogue watch face allows us to find direction. In the Northern Hemisphere this is achieved by the following method. Check that your watch is accurately set to local time, reset for any local summer time which may have been added.

- Point the hour hand at the sun.
- Using a thin twig, cast a shadow along the hour hand through the central pivot.
- Bisect the angle between the hour hand and 12 o'clock.
- This line will be pointing due South, North being furthest from the sun.

The same procedure applies in the Southern Hemisphere, once again having set your watch to local time:

Shadow the sun along the hour hand and through the central pivot to determine direction in the northern hemisphere.

Shadow the sun along the 12 o clock position and the central pivot in the southern hemisphere.

Stick and Stone Method

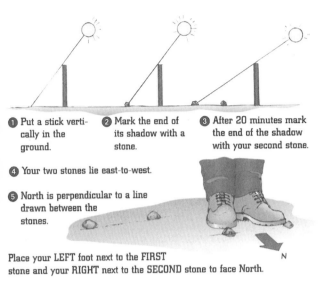

1 Put a stick vertically in the ground.

2 Mark the end of its shadow with a stone.

3 After 20 minutes mark the end of the shadow with your second stone.

4 Your two stones lie east-to-west.

5 North is perpendicular to a line drawn between the stones.

Place your LEFT foot next to the FIRST stone and your RIGHT next to the SECOND stone to face North.

- Point the number 12 at the sun.
- Using a thin twig, cast a shadow to achieve more accuracy.
- Bisect the angle between the hour hand and 12 o'clock.
- The end of this line nearest to the sun indicates North.

Navigation by Night

Navigation by the stars has been practised for thousands of years, and having learnt about the stars is hugely beneficial in survival navigation. Bright stars that seem to be grouped together in a pattern are called constellations. The shapes of these constellations and their relationships to each other do not alter. Because of the Earth's rotation, the whole of the night sky appears to revolve around one central point; and this can help you to find directions.

The North Star In the Northern Hemisphere a faint star called Polaris, the Pole or North Star, marks the central point. Because of its position it always appears to remain in the same place – above the North Pole.

As long as Polaris can be seen the direction of True North can be found. To find Polaris, first locate the constellation variously known as 'The Great Bear', 'The Plough' or 'The Big Dipper'. A line through the two stars furthest from the 'handle' always points towards Polaris. Take the distance between the two stars and then follow the line straight for about six times the distance. At this point you will see the Pole Star. If you are unsure which way to look or wish to confirm that you have found Polaris, look for another constellation called Cassiopeia. The five stars which make up this constellation are patterned in the shape of a slightly flattened 'W'. Cassiopeia is positioned almost opposite the Plough, and Polaris can be found midway between them. As long as the sky is clear the Plough, Cassiopeia and Polaris remain visible all night when seen from any country north of 40 degrees N latitude.

The Southern Hemisphere The Southern Hemisphere does not have a version of Polaris conveniently marking the direction of South. Instead you will need to locate the constellation of the Southern Cross, made up of four main stars with a fainter fifth one just off the centre point of the cross. Take a line through the longer of the cross's arms and extend it for four and a half times its length. If you have the right line, it should pass through a group of four very faint stars shortly after the Cross. This will take you to the point where you will find South. To make navigation easier, find a landmark directly below this point to indicate South in a terrestrial plane.

Star Movement Method
Clear skies cannot always be guaranteed, but if there is only partial obstruction by cloud and you are still able to see individual constellations, navigation by the stars can still be employed. The Star Movement Method is based on knowing how the stars wheel around the sky. Depending on which way they are moving, you should be able to get a rough indication of the direction you are facing. To do this you will need two fixed reference points, such as two sticks set in the ground like the sights of a gun. These should be aimed at any prominent star.

Star Movement

If the star moves right, you are facing South; left=North; down=West; up=East

If the star appears to be:

- Looping flatly towards the right, you are facing approximately South.
- Looping flatly towards the left, you are facing approximately North.
- Rising, you are facing approximately East.
- Descending, you are facing approximately West.

You can also use your eye as the second fixed reference point, as long as your head is steadied against some solid object first. With this method, it is best to use a series of glances to observe the star. Fixed staring will produce an optical illusion of either the star wandering about, or not moving at all.

The Moon

The movement of the moon also follows a set pattern and can aid navigation. There are two methods which can be used:

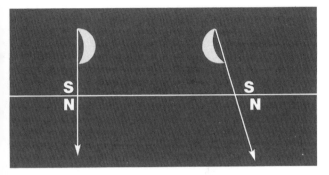

The Quarter Moons Both waxing and waning quarter moons can be used. Draw a line through the horns down to the horizon. The point where it touches gives a rough indication of South, if you are in the Northern Hemisphere, and North if you are in the Southern Hemisphere. Although not very accurate this will at least provide you with a rough guide while travelling at night.

The Moon and Time Make sure that your watch is correct and set to local time. The phases of the moon pass through certain directions at certain times; so by using a watch in conjunction with the table below you should be able to get a good idea of the direction you are travelling in.

	1ST QUARTER	FULL MOON	LAST QUARTER
1800	South	East	
2100	Southwest	Southeast	
Midnight	West	South	East
0300	Southwest	Southeast	
0600	West	South	

Vegetation Tips to Navigation

Using vegetation to find your direction is not very accurate, but can give you an idea of your heading. Trees and shrubs which are perennial tend to produce more foliage on the side most exposed to the sun, that is the side nearest to the Equator.

Strong prevailing winds may also cause a greater amount of foliage to grow on the leeward side. Use common sense and observation to discern whether wind or sun has had greater effect. Moss prefers the damp and the shade, tending to grow on the side of a tree or rock that is least exposed to the sun – i.e. they tend to grow facing towards the Poles rather than the Equator.

Your decision to stay at the crash site or walk out will need careful consideration. If your chances of location seem poor, walking out may be your best option – although it might take months, and demand the development of long-term survival skills.

The factors governing any decision should be based on where you are, your chances of survival if you stay there, where you intend travelling to, and the related hazards in getting there. Low-lying mist can frequently hamper aerial searches of jungle terrain. If your site is clearly visible from above, and you have workable signalling devices, you may choose to stay.

A realistic judgement of your physical, mental and material resources – your ability to travel and to reach a given point accurately – must be made if you choose to move. Do you have casualties? How many, and how badly injured? It takes a minimum of four healthy survivors to carry one injured person through the jungle, and even with this ratio the journey will be long and hard.

Once you have determined the need to travel you must prepare. Before you start, it is important to check the weather and work out a travel routine based on the type of terrain and conditions you will be passing through.

The pace should be steady and unrushed, with a break of five to ten minutes at least every hour. Use this break productively, not just to rest but also to evaluate your progress so far and to consider the next part of your route. This is also the time for minor repairs and adjustments to clothing and kit. Make sure that you do not go beyond your physical limits – and take into consideration that your feet will be doing

Travel Equipment

① Wrap your kit in a blanket or tarpaulin and carry it over your shoulder.

② Alternatively, make a pack to carry your kit on your back.

LIFESAVER
Obstacles

Where survival travel is concerned, obstacles can be divided into two types: natural features, and living creatures. Natural obstacles obviously include e.g. rivers, swamps and mountains; while wolves, bears, elephant, rhino, crocodiles and alligators are but a few of the dangerous animals a survivor may face.

most of the work while you are travelling. It is wise to take care of them; prevention is much better than cure. At the end of every day's march remove all footwear and wash your feet; also wash and clean your socks, stockings or footrags, and boots or shoes. Make a fire and dry the footwear overnight; that way it will be clean and dry for you to put on again the next morning.

Horseshoe Bandoleer This is one of the easiest packs to make, and can be carried quite comfortably over one shoulder. To construct it you will need a square-shaped piece of material, such as a blanket. This should be laid flat upon the ground and the items to be carried placed along one edge – and padded, if necessary. The items should then be rolled within the material towards the opposite edge to form a sausage shape. The two open ends need to be tied securely, and more ties added along the length to prevent the contents shifting. The two ends can then be joined with another piece of soft material, and the pack is slung round the body.

Square Pack For this pack you will need some sort of rope or cordage. Hopefully this is already available, otherwise you may need to make some from whatever materials you have about you. Once you have this, you will be able to construct a square or wishbone frame from sticks or bamboo.

Route Selection

If you can, choose to follow a trail along a ridge rather than a route that takes you through a valley. Valley routes generally present more obstacles, such as thick undergrowth and possible river/stream crossings. Other hazards include swamps or marshy ground, which are at the best hard going and at worst, dangerous to navigate. If you have alternative routes, it is always best to detour around such areas.

Ridges also tend to provide better visibility, which will make it easier to keep your bearings. A ridge may be orientated in the direction you are travelling; more often than not, however, it is likely that the direction of the ridge will head off on a totally different bearing to the one you wish to follow. Even so, it may still be worth following the ridge for a short while, keeping an eye out for a suitable alternative route to take.

Contouring offers a useful halfway measure between ridge and valley floor. A trail that follows a contour may take a longer route than a ridge top, but it will mean that less climbing has to be done. Without a detailed map your route selection is best made based on careful observation of the terrain.

Don't Fight the Jungle

• Don't try to cut a path to follow a compass bearing rigidly.

• Seek areas of thinner vegetation, game trails, streams, etc, which lead roughly in the right direction.

• Follow ridge tops if possible.

• Move slowly, stopping for frequent compass checks.

• Don't travel after sunset.

Obstacles & Hazards

Dense Vegetation The jungle presents the survivor with an obstacle course; vegetation tends to be very dense, and cutting a path will be slow and exhausting. When travelling through this type of terrain normal progress will be about 1km (0.6 mile) per hour, or about 5km (3 miles) per day. It is therefore vital to choose your route following the largest opening in the vegetation that is roughly in line with your intended direction. Don't just look – penetrate the forest with your eyes. Move slowly, stopping regularly to orientate yourself. Native paths, game trails, dry water courses, rivers and streams or ridge crests all offer a slightly easier passage. Bear in mind, however, that animals will also use these trails, especially at night – after sunset it is safest to stay in camp. Take precautions to limit the discomfort and damage caused by insects and leeches; and always check for wildlife in your bedding, packs, clothes and especially boots before use.

Swamps Low-lying jungle areas near to water or where the ground is poorly drained will have swamps. There are two types: mangrove or saltwater swamps, and palm or freshwater swamps. Mangrove swamps occur on coastlines subject to tidal flooding. Mangroves thrive in these conditions; they are small trees, usually from 1m-5m (3ft-16ft) high, although some can grow as high as 12m (40 feet). They have extensive, tangled root systems both above and below the water which can create hazards for the traveller, both on foot and on a raft. Due to the dense growth visibility is poor both on the ground and from above. These swamps may also harbour large crocodiles, leeches and biting insects. Despite the abundance of food they should be avoided, or crossed as quickly as possible.

'Freshwater' or palm swamps actually occur in saltwater as well as in freshwater areas. Undergrowth usually consists of small palms, thorny bushes, reeds and grasses, which can make movement and visibility extremely difficult; however, they often have channels navigable by a small boat or raft. This sort of swamp has much wildlife of survival value and fewer animal dangers than the mangrove swamp.

Savannah On these wide, open stretches of grassland characteristic of much of Africa small shrubs may be found but tall trees are rare. The thick, broad-bladed, sharp-edged grass can grow from 1m to 5m (3ft-16ft) tall. Depending on its height, movement may be slow and tedious, and there is little shade from the sun. Visibility may be poor at ground level, but is reasonably good from search aircraft.

River Crossing Techniques

Sometimes when a survivor is confronted by a major watercourse there is little option other than to cross it. The width of the river, its depth and speed all pose problems, as does the type of riverbed.

Mud and silt can be extremely dangerous, to the point where you become permanently stuck or – worse – sucked below the surface. Few riverbeds are flat, and most have hidden depths into which anyone crossing may fall. Strong currents can dash the strongest of swimmers into rocky outcrops, or plunge them into falling rapids. All these obstacles must be taken into account before attempting to ford any river.

Quicksand

Should you fall into quicksand the danger you face is determined by its consistency and depth. Despite the movie version, most quicksand holes are very small; it should be possible to reach firm ground, or grab a purchase which enables you to pull yourself free. If you cannot do this, don't panic:

• Immediately throw yourself prone, and 'swim' using the breast stroke.

• In very wet quicksand you may go under, as you would in water; but the retention properties of wet quicksand are weak, and you should be able to surface and swim.

• Stagnant quicksand is more dangerous, as it will grip your body and slowly suck you down until you reach the bottom of the hole. In this case any abrupt motion may dig you deeper.

Never cross on a river bend, as the speed of the current and the depth of the water will increase from the inside of the bend to the outside. Instead, choose a wide stretch where the water is flowing slower, and where one can carefully wade across. Avoid any temptation to jump from stone to stone, as these are often slippery and a fall could result in a sprain or a fracture; at the least you may drop vital equipment. When wading across, use as many aids to safety as are available.

If you flounder or slip in the water and find yourself floating downstream, it is important not to panic. Float feet first with the current, fending off any obstacles, until you feel the river bed beneath you and are able to stand, or until you reach the safety of the bank.

Depending on the weather conditions you are advised to remove your socks and trousers before wading any river; this will allow you some increase in comfort and warmth when you dress after reaching the opposite bank. When crossing a river alone use a stout stick to provide extra stability in the water and to test ahead of you for depth, potholes and underwater obstacles.

If a rope or line is at hand and you are with a companion, make sure that the person crossing is safely secured to the bank. A survival party of three or more should follow the procedure illustrated here, which will ensure that the person crossing is always secured by at least two others.

If the water is too deep for wading and you have to swim across, make buoyancy aids. Your rucksack secured in an airtight survival bag will make an excellent floatation aid. Injured survivors should be assisted by the party linking arms around each other's shoulders, with the weakest swimmer in the middle. If necessary, place the casualty on top of a secured bouyancy aid such as several rucksacks lashed together. Move across the river with the strongest member on the upstream side, against the flow of the current; move slowly and support each other. Take care when entering and leaving the stream, especially if the banks are steep; hold onto the bank, and help the weakest person out first.

3-Man River Crossing

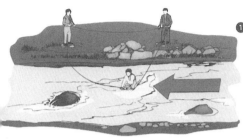

1. Make your rope into a loop. The first man can cross, inside the loop for extra safety, supported by the other two.

2. The second man can cross the river supported from both banks of the river.

3. The third man can then be helped across by the first two.

Rafts

Traversing jungle terrain by river is still one of the most popular methods of travel, due to its speed; but for the survivor it presents the twin threats of dangerous animals and dangerous water. Rafts are best built from bamboo lashed together with vines – these materials are plentiful and easy to work with.

Forget any ideas about constructing a canoe - it takes skill and a great deal of time, and requires special resources such as resin for waterproofing. However, making a raft is within everyone's capabilities. Rafts can be built with anything that has a degree of

River Tips

• Plan your route to avoid having to cross water if possible. Cross only if absolutely necessary.

• Always look for the safest crossing place. Choose the widest and shallowest stretch, and avoid bends.

• Unless in a life-threatening situation, never attempt to cross a river in flood.

• If alone, use a buoyancy aid.

• Use a safety line.

Floatation Aids

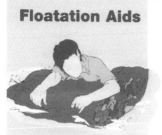

buoyancy – you do not necessarily need the luxury of old oil drums, or even timber; empty bottles tied up inside a polythene bag will serve you well. If using wood, choose a light wood such as bamboo if available.

1 Foliage wrapped in tarpaulin will make an efficient floatation aid.

Once your raft has been constructed, ensure that all your supplies and equipment are firmly secured to it in case the raft tips over. Do not attach too much weight to your person – if you fall in it may prevent you from staying afloat. To steer your craft in order to avoid any river obstacles such as rocks or rapids, use a long punting pole. Leave time at the end of each day to locate a river-bank campsite. Mosquitoes and other biting insects are liable to be a hazard close to water, so locate your camp on higher ground if possible. A higher location will also prevent problems if the river level rises quickly. This can happen without warning in a rainforest, even if there has been no rain in your locality, as rain may have fallen in the upper reaches. When you set up camp for the night, take great care that you have secured your raft, even taking it out of the water if possible. Check the raft for serviceability every morning and repair as required.

2 Cut two lengths of good sized timber and lash them together to make an 'armchair'.

A Poncho Raft

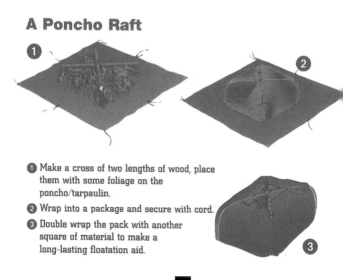

1 Make a cross of two lengths of wood, place them with some foliage on the poncho/tarpaulin.

2 Wrap into a package and secure with cord.

3 Double wrap the pack with another square of material to make a long-lasting floatation aid.

Constructing a Raft A raft is basically anything that floats with enough buoyancy to support a human being. The secret behind any good raft is its strength and durability over water; it does not have to keep your feet dry. The materials for making a raft can vary from brushwood and logs to polythene sacks and ponchos. Their construction needs little imagination and several ideas are illustrated here. Keep in mind that it is trapped air which makes a raft float.

Mountain Travel

Mountains, especially the lower slopes, offer the survivor or escapee many possibilities of shelter, food and water. Their drawbacks are that travelling may be more difficult, and the survivor

Climbing with Rope

If you have a rope use it for both climbing and descending, it will make your travel through mountains far safer.

Climbing without Rope

If you have no rope always maintain three points of contact when climbing. Do not overstretch yourself. Face away from the rock when descending.

may be at more risk of injury. Remember when travelling through mountains:

1. Avoid any loose rock when climbing, and always make sure that you have three points of contact.
2. Always make sure that it is possible to climb back down again if you need to.

Movement in the Dark

Night travel should not be undertaken in the jungle but there may be times when a survivor needs to move during the hours of darkness. In the jungle this is always best avoided and may often be completely impossible but it may be expedient to move by night to avoid the enemy, or to get a seriously injured survivor to immediate medical attention. If, for whatever reason, moving in darkness is the only option in your situation, you need to consider how to do this safely.

Being in complete darkness can be frightening; stay calm and take stock of the situation. Check that you have no source of lighting on you. If you are moving in a group, make sure that everyone stays within touching distance of the next person, placing the weakest in the centre.

Unless a life really depends on it, do not try crossing a river in darkness – it is extremely dangerous. Likewise, while it is a good idea to follow a stream or river on the flat, never follow water down a steep mountainside, as it will inevitably have a waterfall somewhere. Even if the waterfall is small, it will be enough to cause injuries if you fall over in the dark.

Working in complete darkness produces what is known as night sight, a condition where the eyes adjust to the low level of available light. This will be interrupted if a torch or naked flame is used; always close one eye against bright lights. All the human senses become heightened when enveloped in darkness, and these should be used to their best advantage. However, be aware that heightened senses mean that even familiar noises may sound much louder and closer, which to some people can be unnerving. Stay calm, and talk to yourself or each other if necessary.

Prior to darkness falling, try to check the ground you intend to cross and memorize your route. If your memory is good it will assist in maintaining your route during the darkness. Distances can be confusing, as you will be forced to move more slowly. Try, if possible, to locate features which can be easily identified. Your sense of touch will be particularly useful when it is totally dark or when you are moving over steep and rocky ground. Again, always move downhill, using your hands as if you were a climber, keeping three points in contact with the rock at all times.

Use your hands and arms to make sure that the immediate space before you is clear of any obstacles and is secure to step on. If the ground is uneven or if there is the possibility of a dangerous drop, crawl on your hands and knees. Stop when you hear water, as this almost certainly indicates a drop of some height. Try throwing a stone and listen for the sound of it hitting earth or water; this should indicate distance and depth. In a forest at night, stretching the arms out in front of the body will ward off low branches, etc.

Keeping the mouth open during darkness increases sound reception. Furthermore, whenever danger is imminent the senses often produce a tingling sensation as a warning. Learn to recognize these signs and accept them without interpretation – remember that we are, after all, only animals.

Local Population

The hazards of surviving in desolate terrain may seem to be over if you make contact with the indigenous people – but this may be far from true. Many travellers have found themselves held hostage for ransom, or have been imprisoned for breaking some fundamental local law. That said, there are few places on Earth where meeting up with the local people will not bring your survival situation to an end, provided that you obey some very basic rules.

Only the most hostile of environments are totally uninhabited, since both man and animals have adapted to virtually every type of terrain and conditions. The nomadic Tuareg have survived in the central Sahara, while the Inuit have eked out an existence in the Far North. Today many tribes have become engulfed by the trappings of Western civilization, but many still cling to their former ways of life. In many cases these isolated peoples will have some form of communication with the outside world; if not, then at least they will have food, water and shelter, and offer a hopeful means of repatriation.

In the event that you make contact with such a community, there are a number of basic rules to observe.

- Unless you are at death's door, wait outside their village until welcomed.

- Lay down any weapon, but do not let them take it.

- Spread out your palms to show you are unarmed before shaking hands.

- Have an escape route planned in case you have to run.

- Take any drink or food that is offered, and thank your host.

- Treat all people, customs and religions with respect.

- Explain your situation – use simple drawings in the dirt.

- Talk to the men; do not openly approach or talk to the women.

- Other than for medical reasons, don't remove clothes in public.

- Explain that you must move on as soon as possible.

- Thank them for their kindness when you leave.

Search and rescue contin-
gency plans come into
operation at the first sign
of trouble. This will happen when a radio distress
call is received, or way-
point call-in procedures
have been missed.

All air traffic is monitored very closely, especially commercial aircraft flying on a set route. Radar and the more recent adoption of the satellite Global Positioning System mean that almost all transport vehicles, from aircraft to super-tankers, can be tracked constantly to within a few metres. Consequently, aircraft that have been forced to make emergency landings or ships that are foundering at sea can usually expect rescue assistance within a matter of hours at worst. If for any reason the location is not known, then Search and Rescue (SAR) teams will be called in to make a search. The area covered will be based on the best estimate of the last known location.

How the search is carried out will be determined by the size of the area to be covered, the terrain, the weather and operational necessity. A search plan will be devised, and search patterns allocated to the aircraft. If radio communications can be established or a beacon signal is received, then a contact search will be initiated. This is designed to concentrate rescue efforts on a relatively small area, thus increasing the speed with which rescuers can get to you. Unless there is accurate knowledge of the location of the party to be rescued, it will be futile and even risky to send out search teams during the night.

Put Yourself In the Searcher's Place

- Making effective signals depends not on what you can see from ground level, but on imagining what SAR crews can see when looking dowwards or obliquely at your location.

- The primary aim is CONTRAST – of colour, shape, movement.

- Against forest, only dense white smoke shows up.

- Always keep signal fires ready to light instantly.

The basic search patterns are as follows:

- **Area search** This involves dividing up the area into smaller areas using natural features as landmarks, giving boundaries in which individual teams are to search.
- **Sweep search** The rescue party will spread out in a line and search the area in a disciplined and organized manner.
- **Contour search** In mountainous country a contour search, spiralling around tall features and flying several times along steep valleys, allows maximum thoroughness.
- **Contact search** A search focused on a smaller area but based on the principles of the sweep search.

Signalling

Radios, rescue beacons, strobes and radar-reflective balloons should be activated as available and required; where none are available, improvise. If you stay at a crash site you must use every possible means to signal any rescue aircraft. The jungle density prevents sound from travelling and the noise of an aircraft will not be heard until it is close. You must therefore display some form of permanent signal; the best is the aircraft or parachute itself, especially if it is still up in the forest canopy.

Signal Fires

Signal fires must be kept ready to be lit at a moment's notice. The tinder and wood must be dry. The fire is constructed so that the tinder is in the middle, ready for lighting. The aim is that once the tinder is alight all the other fuel should light easily and burn without too much effort on your part. This type of fire needs to be sheltered from the wet. Using any oil or petrol that you may have can speed up ignition.

The canopy will hide and disperse the smoke from the largest fire (and if the surrounding vegetation catches fire you are liable to have more immediate concerns than your rescue). Ground signals can obviously be seen best where there is a break in the canopy, and any rescue aircraft will be checking every clearing. Rivers and their banks are normally visible from the sky. Jungle rivers are often pitted with small islands large enough to build a fire on; failing this, make a fire platform and tether it midstream. If a fire is not feasible use marker panels instead.

Your rescuers may appear at any time, so when the moment arrives remember that you need to produce contrast; to work properly the signal fire must stand out from its background. At night it should produce tall, bright flames which are easily seen but in the daytime you will need to produce more smoke than flame.

Tree Signal Fire In many cases it will be possible to construct a signal fire using a single growing tree. Use one that is isolated or growing on the fringe of a clearing. Make sure that when it is set alight the fire will not spread to the surrounding trees; clear smaller trees to isolate it if necessary. Thicken your signal tree by stripping branches from other nearby trees and interweaving them. Next build a small fire under the base of the tree; this will act as a booster, and will be partially shielded by the wide branches at the base. This booster fire will need to ignite immediately; either have a container of combustible fuel handy, or – if available – use a

Personal Locators

There are many devices used for contacting and locating those who have become lost, and most have similar functions and operations. One example is the SARBE 6 (Search And Rescue BEacon), which is designed for use as a survival radio by civil or military aircrew. On activation the unit transmits a continuous, internationally recognized, swept-tone radio distress signal in the UHF 243Mhz 7.5kHz or VHF 121.5Mhz 3.75kHz distress frequencies. It also provides two-way voice communications between the survivor and approaching rescuers. Built-in self-test facilities allow a simple confidence check to be carried out for correct functioning of the unit and battery state.

The unit is activated by the removal of an operating pin, either manually or automatically by such functions as liferaft inflation or ejector seat operation. Simultaneous, omnidirectional transmission of both VHF and UHF signals then continues automatically for a minimum of 24 hours to facilitate detection by search aircraft or vessels or by any other land, sea or airborne installation monitoring these frequencies. Pressel switches located on the side of the unit allow the survivor to select the voice mode, permitting two-way communication with the rescuers. Voice communication is on both distress frequencies simultaneously. This mode is intended for use only when the survivor can see or hear the rescue craft. The SARBE 6 is waterproof to a depth of 10m (33 feet).

salvaged aerosol can (hair spray is excellent). Your signal fire tree should be protected from the rain if at all possible by covering it with a parachute or similar canvas. If nothing is available, check it daily and shake it to detach fallen water.

Pyramid fire A pyramid fire needs a raised base and plenty of dry fuel. The aim of the base is to ensure quick ignition and a good air supply once alight. As with any signal fire, the fuel should be instantly combustible and stacked in a manner that allows air to permeate and feed the flames.

If you have a suitable salvaged aerosol can, half-bury this in the ground at an angle pointing upwards at the pyramid base. Have ready at hand a large flat stone heavy enough to depress the release button, and a torch made from a length of stick with rags wrapped around one end. At the first sign of rescue aircraft light the torch, put the stone on the button, and place the torch in the spray. Turn your back on the fire when you do this. Once the fire is burning well, distance yourself until the spray can is finished or has exploded.

Warning: Setting fire to aerosol spray is highly dangerous, and should only be attempted in a dire emergency.

Even in a survival situation extreme caution is advised, as the can will almost certainly explode.

Phones Although mainly restricted to land usage, the global

A heliograph can be purpose-made or improvised from a vehicle mirror.

telephone network is extensive and accessible in many remote places. New portable satellite phones are little larger than a laptop computer, and will operate in every environment. Anyone planning to travel or spend any time in regions where survival situations might occur should investigate beforehand access to all forms of telecommunications, from land lines to mobile phones.

Searching the passengers and luggage from any wrecked aircraft will produce a variety of communication equipment, which even if not operable from the present position may connect later on.

Mirrors and Heliographs Any type of mirror – the larger the better – is excellent for signalling providing you have bright sunshine. It is simply a matter of reflecting the sun's rays towards a search plane or party to attract their attention. All aircraft or vehicles carry a number of mirrors any one of which will serve as a signalling device. A more accurate method is to use a purpose-made heliograph. Modern variations of these are smaller than a computer disk, measuring just 5cm x 5cm (2in x 2in), yet they have the capacity to accurately reflect some 85% of sunlight up to a range of some 20 kilometres (12 miles). Once any rescue aircraft gives definite signs of having spotted you, stop signalling – you will only dazzle the pilot.

Light Light is obviously the ideal means of attracting attention at night, even after you have made radio contact. Light can be emitted from any number of sources: a naked flame, torch, strobe, camera, or flare.

Although they are extremely effective the problem with most flares and torches is that they are either limited to a single use, or are useful only for the duration of the batteries. All survival flares come with operating and safety instructions; make sure you read these before commencing any operation. Hand-held flares might be better used to ignite a larger signal fire.

Parachute and Missile Flares
There are many different types of missile flare on the market. Some simply fire a glowing light which lasts a few seconds; some have a parachute attached, which will retard the flare's descent thus making it visible for longer. Always read the instructions carefully and follow them to the letter. The important point is always to keep the flare pointing skywards. Parachute flares are one-shot devices, so make sure their use is justified. The number of flares supplied with any normal pistol is around nine.

Torches and Strobes Any torch is a bonus at night, but for signalling purposes a large, broad beam is required if any rescue aircraft is to see it. Moving the torch from left to right in a slow arc will help attract attention, as will shining it onto a reflective surface – it is not the light which the search aircraft crew will see, but the movement of light. Strobes are designed to create this effect by emitting an extremely bright pulsating light. On a clear night a strobe can be seen some 16km (10 miles) away.

Vehicle Lights Providing certain elements are still intact a good signalling light can also be generated from vehicle and aircraft lights. These lights are best aimed at a large surface, such as aircraft wreckage, with the light being fanned to animate movement. If done properly this will create an effect that can be seen for many miles. Remember that the system is reliant on battery power and should be used sparingly to save the batteries.

Camera Flash A modern camera flash also makes a good signalling device but, as with other battery-powered systems, it has a limited life and should be used sparingly.

Whistles and Sound The main purpose of the whistle in many survival kits is to attract the

Personal Locator Radios

• This is the survivor's best friend - but its battery life is limited. The 'beacon' setting uses least battery power.

• Reception and transmission are generally limited to line-of-sight – so don't waste the battery by leaving it switched on for long periods when you can't see or hear SAR aircraft.

• Try to use the radio from high ground.

• Transmit an SOS or Mayday at sunrise, noon and sunset. Try to transmit during a consistent time period, e.g. from the hour to 20 minutes past the hour.

• Transmit for two minutes each time, then switch off for one minute. Switch on for three minutes, then off for three minutes. Switch on for ten minutes, then off until the next transmission period. (USAF guidelines for use of standard PRC-90 radio.)

All pyrotechnics are dangerous, make sure you read and conform to the instructions for use.

attention of other survivors directly after the disaster. This is particularly so at sea, where all survivors should find a whistle attached to their life vest. On land anything that will amplify sound, such as beating a metal drum with a stick, should also be considered. If the survivor is lucky enough to have a firearm, firing a shot will also attract attention; but this should only be done if you believe a rescue party is nearby.

Any form of signalling reliant on sound will be rendered far less effective in jungle conditions due to the density of the vegetation.

Balloons Radar-reflective and colour-detectable balloons come in a variety of sizes. The coloured versions are primarily designed for use in the jungle. Normally constructed of bright orange polythene, they are inflated by mixing chemicals with water to produce helium gas. As the balloon fills it is raised on a line until it is above the dense forest canopy, and then tethered where it can be clearly seen by search aircraft. (The water bag and tin which are part of the kit may be employed for survival purposes.)

Ground marker panels.

Shapes for Specific Ground Signals

N	Negative	→	Have gone this way
Y	Yes	△	It is safe to land
I	Have seriously		here
	injured	**SOS**	Save our souls
X	Unable to move		

Radar-reflective balloons are more compact, and are more auto-mated in their operation. Inflation is initiated by removing a safety pin; this activates a helium cartridge which fills the balloon. The balloon is tethered to the life jacket, from where it rises to around 30m (100ft), where it will remain even in strong winds for up to five days. The 10m (33ft) radar reflective signature can be detected by search vehicles up to 30km (18 miles) away.

Rescue Panels and Streamers These come in various shapes and sizes, but all provide a fluorescent marker which can be seen from the air. Panels are normally 2m x 0.5m (6.5ft x 16ins) or 2m square depending on the design. Two or three of these can be formed into various shapes which indicate your requirements and situation. Distress streamers are used in much the same way but are narrower and longer, up to 10m (33 feet). These can be spread on the ground in a clearing or on top of the jungle canopy if you are able to reach it.

Contrast Signals Disrupting the normal pattern of the terrain creates contrast. Do this by introducing regular shapes which do not occur naturally – circles, squares, triangles, letters or straight lines. A large circle with a minimum diameter of 3m (10ft) can be made using stones. It can also be broken or trampled-out in the grass of a clearing with the addition of some contrasting material – earth, campfire ashes, even marker dye if a dinghy is found among the wreckage. Choose the things which make the best contrast against the particular background surface. If air marker panels are available use these first and construct improvised signals secondly. Make any ground-to-air signals as large as pos-sible, and add extra shapes if space, time and energy permit.

Increased contrast is gained if you incorporate brightly coloured wreckage, clothing, blankets, etc, in your signals.

Jungle Extraction

Jungle survivors unable to walk or float back to civilization will usually depend on helicopters for any rescue. (Ground searches may be organized, but in tropical forest the chances of locating even something the size of an aircraft are not high.) Helicopter rescue also has its pitfalls, due to the denseness of the forest canopy. This makes locating survivors extremely difficult; and the task of extracting them from the jungle floor presents special problems.

Flying Signals Signals can also be hung from trees. Anything shiny or brightly coloured which is moving will be even more eye-catching. A flag pole will increase the distance over which signals can be seen from the ground. If any possible rescuers are seen or heard, use any available clothing or material as flags, and keep waving. If there is some suitable material that is not required for other uses during daylight hours, it is useful to have it ready, attached to the longest pole you can easily handle for the sake of maximum signalling movement. Two men holding a survival blanket, flag or other brightly coloured sheet can, by keeping it taut, manipulate it to show flashes of light or colour. These will catch a searcher's eye more readily than the display of a static sheet.

Ground Information Markers If you move from your location you may need to blaze a trail or leave ground markers to indicate direction of travel. It is easy to get lost or move off course, especially if no compass or map are available. To aid your progress and to make

Rescue teams are highly skilled at extracting survivors from the most inhospitable of places.

sure others can follow in your foot-steps you will need to blaze a trail. This can be done either by chipping markers on tree trunks or leaving a prominent ground sign.

Trees should be cut at head height on both sides, making a single cut on the side pointing away from your last position, and two cuts on the side pointing towards your last position. This will allow others to follow and you to retrace your footsteps if the need arises. Always look back from time to time, making sure that your spacing between marked trees allows the next one to be seen from the position of the last.

A second method is to deliberately place natural items such as stones, sticks, grass, etc in such a way as to mark your direction. The distance apart is determined by the natural path you are taking. If you are on a prominent path you need only mark direction changes at junctions.

Helicopter Rescue Procedures

Most SAR teams are organized along military lines. They are highly skilled and have access to

Helicopter Rescue

• Wait until the helicopter has landed, and either the pilot or a crew member has clearly indicated to you that you should come forward.

• Never approach a helicopter from the rear, or by descending down a slope – both will put you in extreme danger from the rotor blades.

• The best approach angle is on the cabin door side, from three-quarter front.

excellent resources, including fixed-wing long range aircraft and helicopters. Most carry personnel and facilities for front line medical care. However, it would be a dangerous mistake to assume that they will always be there to get you out of danger.

Severe weather conditions can keep search aircraft grounded for hours or days. Even once you have been located helicopter crews will take time assessing the problems of trying to reach you. Over jungle terrain it is not uncommon for the pilot to make several attempts to establish a hover close enough to the casualties to be able to get a winchman or mountain rescue team to their position. Having arrived at a workable hover, the next priority is to assess the safest method of rescuing the survivors.

To ensure that no important aspect of the situation is overlooked SAR crews use a standardized system of priorities:

- Aircraft safety
- Winchman safety
- Survivor safety

Landing Areas Where possible the helicopter will land to evacuate survivors. To make this viable the survivor should do everything possible to provide a good landing pad (LP). Factors to be considered include the size of the clear area, the ground slope, the type of surface, and the direction of wind and approach. First check that the surface will support a helicopter, i.e. that it is not waterlogged ground or obstructed by large rocks, fallen trees, potholes, etc. Next, make sure that it is free of any loose debris that could be blown about by the rotor downdraft. Check the helicopter's approach path, which will be into the prevailing wind; make sure there are no tall obstructions to the rotor blades. Mark the centre of your LP with some form of marker such as an H-shape; and indicate the wind direction by improvising a wind sock or making smoke.

Rescue Strop The helicopter rescue strop is designed to facilitate the rescue of survivors. It can be used at sea or on land to lift uninjured survivors of any size with relative ease. The strop is manufactured from nylon webbing. A 'D'-ring is incorporated at each end of the strop; the centre portion is cushioned with a rubber sheet comfort pad tapered at each end and covered with polyester fabric. A sliding woggle through which the two ends of the strop pass enables the wearer to draw the strop close to his body before the ascent. A webbing handle is situated centrally on the strop at the wearer's back, to enable the winch operator in the aircraft to grasp the survivor and guide him back into the cabin.

Winching Techniques Most helicopter rescues involve lowering a crew member to assist those being winched aboard. During this double lift the survivor will be secured by the winchman and they are raised together. In certain circumstances this may not be feasible, and a one-man lift will be organized. When the helicopter is positioned into the wind a rescue strop will be lowered to the survivor, who must be conscious and uninjured. The survivor places the strop over his head with the winch cable to the front. It is adjusted under the armpits by tightening the webbing ring woggle, before signalling that he is ready.

Stretcher cases will always be supervised by a lowered crew member.

If you find yourself in a survival situation, whether by design or by accident, remember

FIVE GOLDEN RULES OF SURVIVAL

1 Life itself is survival, all that changes is the environment or the conditions under which you live. When crossing a wilderness area or entering a dangerous environment do so properly equipped.

2 A quick rescue is the best rescue. Use every modern aid to make others aware of your plight and location. No matter what the danger, good communications will prevent a survival situation evolving.

3 We only need the basics. Given that you are uninjured and functioning properly, you need only air to breathe, water to drink, food to eat and shelter from the environment. However scarce, nature supplies all these elements but you cannot expect nature to change in order to accommodate your requirements. You must learn to adapt to use whatever she provides.

4 Plan your survival. When disaster strikes think about your situation and make a plan of your basic needs. Through the practical applications of survival you will maintain hope and give hope to others. Maintain your health, care for the injured, build a fire if you are cold, eat if hungry and sleep when tired. Do not needlessly expend energy or put yourself in danger without good reason.

5 Recognize that danger is everywhere. The cold can kill. The heat can kill. The sea can kill. Wild beasts can kill. Despondency can kill. Lack of nourishment can kill. Watch, listen, think and determine the problem – learn to survive.

Acknowledgments
The information contained in this book was researched from the world's elite surviva
schools, both civilian and military. The use of photography as opposed to illustration
has been adopted to indicate that the processes and methods actually work. Althou
the photographs are from various sources world-wide the contributions of Guy
Croisiaux and Neil Devine are especially outstanding. The visit and introduction by E
Gen. Dato Modh Azumi bin Mohamed, to the survival school in Malacca was very m
appreciated, supplying in the process some outstanding jungle photography. As alw
without the skilful work of Julie Pembridge in coordinating the written material and h
ing this book develop, attainment would not have been possible. Finally the author
would like to express his gratitude to the CEO of BCB International Ltd, Andrew Ho
through whose office many of the contributing companies and individuals were cont
ed for their support.

BCB International Ltd. UK
Beaufort. UK
Special Forces Survival School Malacca.
  Malaysia
International Long Range Patrol School.
  Germany
The Survival School. Belgium

The Royal Navy Survival School. UH
Director of Public Communication,
  Pentagon. USA
Manchester Airport Security Training
Martin Baker. UK
GQ Parachutes. UK
Marinair Holdings. Malaysia